I0066828

James W. Queen

Photographic illustrations for projection, plain or colored

photographic views from nature in all parts of the world

James W. Queen

Photographic illustrations for projection, plain or colored
photographic views from nature in all parts of the world

ISBN/EAN: 9783741162725

Manufactured in Europe, USA, Canada, Australia, Japa

Cover: Foto ©berggeist007 / pixelio.de

Manufactured and distributed by brebook publishing software
(www.brebook.com)

James W. Queen

Photographic illustrations for projection, plain or colored

Photographic Illustrations

FOR PROJECTION,

PLAIN OR COLORED.

3 Inches Diameter, $1.50.   3 ½ Inches Diameter, $1.75.

Square, $2.00.   Uncolored, 50 Cents.

PHOTOGRAPHIC VIEWS FROM NATURE

IN ALL PARTS OF THE WORLD.

EXTENSIVE COLLECTION

ASTRONOMY,	PHYSICS,
LIGHT AND OPTICS,	BOTANY,
ANATOMY,	GEOLOGY,
CRYSTALLOGRAPHY,	GEOGRAPHY.

1897

Entered according to Act of Congress, in the year 1891. by JAMES W. QUEEN & Co., in the Office of the Librarian of Congress, Washington.

NOTICE

Having the largest and best assorted Stock of Mathematical, Optical, and Philosophical Instruments, both of foreign and domestic manufacture, in the United States, we are enabled to offer unequalled facilities and inducements to intending purchasers.

In ordering Instruments and materials from this Catalogue, it is merely necessary to state the edition and the trade numbers of the articles. *All former editions are superseded by this one.*

All Instruments and materials sold by us are warranted perfect for the purposes intended ; and if not found so upon receipt, should be returned and exchanged for others.

TERMS CASH.

The prices throughout the Catalogue will be strictly adhered to.

When no satisfactory Philadelphia or New York reference is given by the party ordering the goods, the money should accompany the order ; but where it does not (either from want of confidence or other cause), the goods will be forwarded by express, with bill, C. O. D. (collect on delivery), *provided a remittance equal to one-third the total amount of the order is sent with it.*

Nor order for a less amount than Five Dollars will be sent C. O. D.

The Express Company's charge for collecting and returning the money on C. O. D. bills must be paid by the party ordering the goods.

The safest and most economical method of remitting money is by Bank Draft or Post-Office Order, made payable to us. Where neither of these can be procured, United States or National Bank Notes, or Postage Stamps, can be sent by express with safety, the sender prepaying the express charges.

Goods ordered to be sent by mail must be prepaid, and the return postage or freight included in the remittance.

Packing-boxes will be charged for at reasonable prices, and all goods will be packed with the utmost care ; *but no responsibility* will be assumed by us, for *breakage* or other *damage, after* a package *leaves our premises,* except upon special contract.

IMPORTING INSTRUMENTS FREE OF DUTY.

By authority of Act of Congress, June 22, 1874, all Colleges, Schools, Literary, Scientific, or Religious Societies of the United States, are permitted to import, free of duty, Books, Charts, Engravings, and Instruments to be used in connection with the educational exercises of the institution for which they are ordered.

We shall be pleased to receive orders for instruments to be imported under this Act, and on application we will give estimates and instructions for proposed orders to be thus imported from any foreign country. The present duty on books and engravings is twenty-five per cent. *ad valorem*, while instruments are assessed according to the component material of greatest value.

QUEEN & CO., Incorporated.

Philadelphia, 1897.

LECTURE SETS.

FINE PHOTOGRAPHIC VIEWS.

UNCOLORED, 50 CENTS EACH.

AMERICA.

WASHINGTON CITY, D. C.

With Reading.

1 Mount Vernon, Virginia
2 Christ Church, Alexandria
3 Old Carlisle House, Alexandria
4 Results of grading the streets
5 Pennsylvania Avenue
6 Capitol, N E
7 Panorama from Capitol
8 Panorama from Capitol
9 Panorama from Capitol
10 Capitol, S W
11 Senate, exterior
12 House of Representatives, exterior
13 Dome of Capitol
14 Senate, interior
15 House of Representatives, interior
16 President taking the Oath of Office
17 White House, interior
18 White House, exterior
19 Inauguration Ceremony
20 Washington Monument
21 Dedication Ceremony
22 State, War, and Navy Department
23 Treasury
24 Corcoran Art Gallery
25 Lafayette Square
26 Columbia University
27 British Legation
28 Georgetown College
29 Panorama from Georgetown College
30 Panorama from Georgetown College
31 Home of the Man who lives in a tree
32 Soldiers' Home
33 Howard University
34 Patent Office
35 Post Office
36 House in which Lincoln died
37 Lincoln Emancipat'n statue
38 Seward's House
39 Garfield Tablet
40 Peace Monument
41 Congressional Cemetery
42 National Museum
43 National Museum, interior
44 Smithsonian Institute
45 Agricultural Department
46 Bureau of Printing and Engraving
47 Fulton Statue, in the Capitol
48 The Potomac
49 Negro Quadrille Party
50 Negro Sermon

NIAGARA.

With Reading. By the Rev. J. Comper Gray.

1 Diagrams—*Upper Diagram:* Geometrical plan of the Lakes, showing their relation to Niagara. *Lower Diagram:* Vertical section of River Niagara, showing the dip of the River between Lake Erie and Lake Ontario
2 Diagram—Geological, showing the cause of the receding of the Falls
3 Plan of our Tour; the spots showing the places whence some of the principal Views were taken
4 View from Monteagle House
5 Suspension Bridge. Railway Track
6 Suspension Bridge. Carriageway under the Railway
7 Within the Carriage-way of the Suspension Bridge
8 View of the Falls from Victoria Point
9 The new Suspension Bridge. Near View
10 The new Suspension Bridge. The Buttresses
11 American Fall, from Clifton House
12 Horseshoe Fall, on leaving Clifton House
13 Horseshoe Fall, from the Custom House
14 Terrapin Tower, on verge of Horseshoe Fall, from Canadian side
15 First Curve of Horseshoe Fall, from Table Rock
16 Second Curve of Horseshoe Fall, from above Table Rock
17 The Rapids
18 Nearer View of American Fall, from Clifton House
19 American Fall, from the Ferry, Canadian side
20 General View of the Falls from the Ferry, American side
21 General View of the Falls, from Point View
22 Bridge over the Rapids, to Goat Island
23 American Fall, from the Hog's Back, Goat Island
24 First View of Terrapin Tower, from Goat Island
25 Second View of Terrapin Tower
26 Third View of Terrapin Tower
27 Fourth View of Terrapin Tower
28 Bridge leading to Terrapin Tower
29 Bridge from first to second of Three Sisters
30 The Rapids and Bursting Wave
31 Snow-wreathed Evergreens
32 Frozen Spray
33 View in Winter from the Hog's Back, American Fall
34 American Fall, through the Snow Arch on Luna Island
35 Snow Arch, nearer View
36 Ice Columns under the Bank
37 Snow effect, Luna Island
38 Falls in Winter, from Point View
39 Below the Falls, Winter
40 Below the Falls, Winter. Nearer View of Ice Cone
41 Map of Niagara, for review of the Tour.

WASHINGTON TO THE NORTHWEST.

With Reading.

1 The Capitol, Washington
2 Baggage Smashers
3 Locomotive
4 Interior Pullman Car
5 City Hall, Baltimore
6 Washington Monument. Baltimore
7 Philadelphia—Public Buildings
8 Philadelphia—Independence Hall
9 Philadelphia—Liberty Bell
10 Philadelphia—Girard College
11 New York—Ferry Boat
12 New York—Panorama
13 New York—Panorama
14 New York—Broadway
15 New York—Brooklyn Bridge
16 New York—Elevated Railway
17 New York—Grand Central Depot
18 Albany—The Capitol

19 Albany—The Senate
20 Albany—The House of Assembly
21 Saratoga—Grand Union Hotel
22 Saratoga—Congress Park
23 Niagara Falls—Panorama
24 Niagara—Horseshoe Fall
25 Niagara—American Fall
26 Niagara—American Fall in Winter
27 Niagara—Rapids and Bridges
28 Detroit—City Hall
29 Detroit—Soldiers' Memorial and Opera House
30 Chicago—State Street
31 Chicago—Lake Steamer
32 Chicago—Post-Office
33 Chicago—Stock Yards
34 St. Paul—General View
35 St. Paul—The Capitol
36 Minnehaha Falls
37 Minneapolis—Bridge Square

38 Minneapolis—Mississippi River
39 Minneapolis—The University
40 Minneapolis—Mr. Washbourne's Residence
41 Montana—Cattle Ranche
42 Montana—Cattle Ranche
43 Ascending the Rocky Mountains
44 Ascending the Rocky Mountains
45 Hayden Valley—Yellowstone Park
46 Old Faithful Geyser—Yellowstone Park
47 Geyser Cave—Yellowstone Park
48 Grand Geyser—Yellowstone Park
49 Mammoth Hot Spring—Yellowstone Park
50 The Great Falls—Yellowstone Park

YELLOWSTONE NATIONAL PARK.

All from Original Negatives.

1 Gardiner Canyon, Entrance to Park.
2 Mammoth Hotel and Stages.
3 Liberty Cap and Mammoth Hotel.
4 Minerva Terrace.
5 Pulpit Terrace.
6 East Entrance to Golden Gate.
7 Golden Gate and Bridge.
8 Obsidian Cliff, Beaver Lake.
9 Norris Geyser Basin.
10 Virginia Cascades.
11 Gibbon Canyon.
12 Gibbon Falls.
13 Mammoth Paint Pots.
14 Fountain Geyser.
15 Excelsior Geyser, from Road.
16 Interior Excelsior Geyser.
17 Excelsior Geyser in Action.
18 Sapphire Pool, Biscuit Basin.

19 Riverside Geyser.
20 Grotto Geyser.
21 Giant Geyser.
22 Crater Oblong Geyser.
23 Punch Bowl.
24 Castle Well and Castle Cone.
25 Castle, Bee-Hive and Old Faithful.
26 Old Faithful Geyser.
27 Crater of Giantess Geyser.
28 Crater of Grand Geyser.
29 Keppler's Cascades.
30 Lone Star Geyser.
31 Shoshone Lake.
32 Hot Spring Cone, Yellowstone Lake.
33 Yellowstone Lake.
34 Hayden Valley.
35 Sulphur Mountain.
36 Rapids above Upper Falls.

37 Upper Falls from Trail.
38 Grand Canyon from Brink.
39 Point Lookout and Great Falls.
40 Inspiration Point.
41 Up the Canyon from Inspiration Point.
42 Down the Canyon from Inspiration Point.
43 Canyon and Falls from Artists' Point.
44 Great Falls from below
45 Great Falls, near view.
46 Petrified Trees, near Yanceys.
47 Tower Falls and Canyon.
48 In Norris Geyser Basin in winter.
49 Foliage near Geysers in winter.
50 Great Falls in winter.

The above collection of Optical Lantern Slides of Wonderland are the finest published. Every slide is guaranteed perfect.

A VISIT TO YOSEMITE VALLEY.

40 Views and Reading.

Wawona Hotel
Grizzley Giant
Log Cabin in Grove
Wawona, entire tree
Wawona, Carriage passing through
Mother of the Forest
Stage passing Alder Creek
Panorama of Yosemite Valley
El Capitan and Bridal Veil

El Capitan from S. W
Bridal Veil Falls
Bridal Veil from below
Cathedral Spires
Sentinel Rock
Three Brothers
Yosemite Falls and Merced River
Yosemite Falls from below
Yosemite Falls, Upper Falls

The Two Domes
North Dome and Washington Column
South Dome and Cloud's Rest
Mirror Lake
Vernal Falls
Nevada Falls, Liberty Cap
Glacier Point
The Domes from Glacier Point

FROM LONDON TO THE FALLS OF NIAGARA.

With Reading. By H. Gore, C. E.

FROM QUEBEC TO THE ROCKY MOUNTAINS.

With Reading.

NEW YORK TO THE WHITE MOUNTAINS.

With Reading.

1 New York Central Station
2 Newhaven, Yale College
3 Newhaven, Yale College Art Museum
4 Newhaven, Peabody Museum
5 Newhaven, Centre Church and Avenue
6 Hartford, State House
7 Hartford, Old State House and City Hall
8 Hartford, Connecticut Mutual Insurance Building
9 Hartford, Mrs. Colt's Residence
10 Hartford, Mark Twain's Residence
11 Worcester, Main Street
12 Worcester, Bigelow and Soldiers' Monuments
13 Boston, Panorama 1
14 Boston, Panorama 2
15 Boston, Panorama 3
16 Boston, Panorama 4
17 Boston, Panorama 5
18 Boston, Old State House
19 Boston, Old State House
20 Boston, Bunker's Hill Monument
21 Boston, Faneuil Hall
22 Boston, Christ Church Tower
23 Boston, Franklin Monument and Old King Chapel
24 Boston State House
25 Boston, Old South Church
26 Boston Custom House
27 Boston Post Office
28 Boston City Hall
29 Boston, Quincy Market
30 Boston, Scollay Square
31 Boston, the Mall, Tremont Street
32 Boston, Washington Street
33 Boston, Tremont Street and Mall
34 Boston, Commonwealth Avenue
35 Boston, Latin Schools
36 Boston, Latin School Volunteer Parade
37 Boston, City Hospital
38 Cambridge, Harvard University
39 Cambridge, Harvard University Memorial Hall
40 Cambridge, Seavers Hall, Harvard University
41 Cambridge, University Soldiers' Memorial
42 Cambridge, Washington Tree
43 Cambridge, Longfellow's House
44 Cambridge, Russell Lowell's House
45 Peabody's Tomb, Harmony Grove Cemetery
46 Portland Harbor
47 Portland City Hall
48 Portland, View from Observatory
49 White Mountains, Trestle Bridge, Notch Valley
50 White Mountains, Buckboard Carriage
51 White Mountains, Littleton and Mount Lafayette
52 White Mountains, Echo Lake
53 White Mountains, Profile House
54 White Mountains, Profile Rock
55 White Mountains, Flume Pool.
56 White Mountains, the Flume
57 White Mountains, the Débris washed through the Flume
58 White Mountains, Mount Washington Railroad
59 White Mountains, Mount Washington Summit, Summer
60 White Mountains, Mount Washington Summit, Winter

ENGLAND.

A TRIP TO LONDON.

With Reading.

1 St. Paul's Cathedral from the River Thames
2 St. Paul's Cathedral from Ludgate Hill
3 St. Paul's Cathedral Nave, looking East
4 St. Paul's Cathedral Choir
5 St. Paul's Cathedral Crypt
6 The General Post Office, St. Martins-le-grand
7 Cheapside
8 The Guildhall
9 The Mansion House
10 The Bank of England
11 The Royal Exchange
12 The London Monument
13 London Bridge
14 The Pool
15 The Tower of London from the River
16 The Tower of London
17 The Thames Embankment
18 St. Thomas's Hospital
19 Lambeth Palace
20 The Houses of Parliament from the Thames
21 The Houses of Parliament looking to Westminster Bridge
22 The Houses of Parliament from below Westminster Bridge
23 The House of Lords
24 The House of Commons
25 Westminster Hall
26 Westminster Abbey
27 Westminster Abbey, West Front
28 Westminster Abbey, Nave
29 Westminster Abbey, the Choir
30 Westminster Abbey, Coronation Chair and Stone of Destiny
31 Westminster Abbey, Poets' Corner
32 Henry VII Chapel, Westminster
33 Henry VII Chapel, Interior
34 Henry VII Chapel, Roof
35 Government Offices, from St. James's Park
36 The Horse Guards
37 The National Gallery
38 Trafalgar Square from National Gallery
39 Seven Dials
40 The New Law Courts
41 Ludgate Circus
42 Holburn Circus
43 Staples' Inn
44 Regent Street
45 Piccadilly Circus
46 Burlington House and Piccadilly
47 St. George's, Hanover Square
48 Buckingham Palace
49 Rotten Row
50 The Albert Memorial

A WALK IN THE "ZOO."

With Reading. By H. Gore, C. E.

Plan of the Garden
1 American Black Bear
2 The Syrian Bear
3 The Polar Bear
4 Dromedary, or one-humped Camel
5 Bactrian, or two-humped Camel
6 Babylonian or Asiatic Lion
7 African Lioness
8 The Royal Tiger
9 Chimpanzee
10 Smoking Monkey
11 Gibbon
12 The Marabout Stork
14 The Cassabara
15 Wild Boar

16 The Wart Hog
17 West African River Hog
18 Collared Peccary
19 Sea Bear
20 Leucoryx
21 The Koba or Sing-Sing
22 Burchell's Zebra
23 Quagga
24 Wild Ass of Abyssinia
25 Syrian Wild Ass
26 Rhea or American Ostrich
27 Emeu
28 Wapiti Deer
29 Gayal
30 Indian Buffalo
31 Cape Buffalo
32 Zebu

33 Brahmin Bull
34 Wolves
35 White or Common Pelican
36 Lima
37 Boa Constrictor
38 Great Kangaroo
39 The Markhoor, or Wild Mountain Boor
40 Indian Elephant
41 African Elephant
42 Indian Rhinoceros
43 Sumatran Rhinoceros
44 Hippopotamus
45 Giraffe
46 The Eland, or Camea
47 The Apteryx
48 Whit-Monday at the "Zoo."

A TRIP DOWN THE RIVER WYE.

With Reading.

1 Hereford Bridge and the Wye
2 Hereford Cathedral—exterior
3 Hereford Cathedral—interior
4 The Wye, at Holme Lacy
5 Old Market House, Ross
6 Ross—Broad Street
7 Wilton Castle, from the River Wye
8 Goodrich Castle—exterior
9 Goodrich Castle—Remains of Banqueting Hall
10 The Wye, at Walford and Kerne Bridge
11 The Wye, at Lydbrook
12 Lydbrook—the Railway Viaduct

13 Whitchurch and the Wye
14 Symond's Yat, from the Wye
15 Little Doward Hill, Symond's Yat
16 Symond's Yat, from the Railway
17 Buckstone Rock, near Monmouth
18 Monmouth—general view
19 Monnow Street, Monmouth
20 Raglan Castle—the Entrance
21 Raglan Castle and the Moat
22 Raglan Castle, showing Bridge and Moat
23 Interior of Courtyard, Raglan Castle
24 Bigswear Bridge, over the Wye

25 Village of Llandogo
26 Village of Tintern
27 Road in Tintern Village
28 Tintern Abbey—exterior, from Southwest
29 Tintern Abbey—interior of Transept
30 Tintern Abbey—interior, looking East
31 Tintern Abbey—the West Window
32 Road under the Wyndcliff, Tintern
33 The Wye, from the Wyndcliff
34 Chepstow Castle, from the Bridge
35 Chepstow Bridge
36 Chepstow—general view

ENGLISH RIVER SCENERY.

With Reading.

1 Thames at London Bridge
2 The Pool below London Bridge
3 Greenwich, from the River
4 St. Thomas's Hospital, Lambeth
5 Moulsey Weir, Hampton
6 Thames, near Pangbourn
7 View at Clivedon
8 Illey Mill, near Oxford
9 Nuneham Bridge, near Oxford
10 Cattle by the River
11 River Fall, near Truro
12 Collecting oak bark in the woods
13 Truro River, from the Hills
14 Flushing, near Falmouth
15 River at Portresco, Cornwall
16 Dartmouth
17 The Mouth of the Dart
18 Old Lighthouse on the Dart

19 View at Lydford Bridge
20 Lynmouth, from the Bridge
21 East Lyn at Watersmeet
22 The Mouth of the Lyn
23 The Avon, from Clifton Downs
24 Clifton Suspension Bridge
25 Tintern Abbey, River Wye
26 The Wye at Tintern Parva
27 Tintern Village
28 The Wye, from Wyndcliff Road
29 The Wye at Symond's Yat
30 Warwick Castle, from the River
31 Memorial Theatre, Stratford-on-Avon
32 The Derwent at Matlock
33 River at High Tor, Matlock
34 Dovedale—View in the Dale
35 Dovedale, from the Heights

36 Worcester, from the Severn
37 Knaresboro and River Nidd
38 The Dropping Well, Knaresboro
39 Dropping Well and Petrefactions
40 The River Nidd and Abbey Road
41 River Ure at Aysgarth
42 Aysgarth—the Lower Falls
43 Aysgarth—the Upper Falls
44 The Wharf—Bolton Woods
45 The Strid—Bolton Woods
46 Stepping-stones on the Wharf
47 Rydal Beck Waterfall
48 Old Brathay Bridge
49 Rustic Bridge over the Brathay
50 Allington Castle on the Medway

TRIP THROUGH YORKSHIRE.

With Reading.

1 York, from Station Hotel
2 York Minster, from S. W
3 York Minster, from N. E
4 York Minster, Nave
5 York Minster Choir
6 St. Mary's Abbey, York
7 The Multangular Tower, York
8 Mickelgate Bar, York
9 Ripon Cathedral, from S. E
10 Ripon Cathedral, West Front
11 Ripon Cathedral, Nave
12 Ripon Cathedral, Choir
13 Fountains Abbey, from the Surprise
14 Fountains Abbey, from W
15 Fountains Abbey, Nave
16 Fountains Abbey, Cloisters
17 Fountains Hall
18 Scarborough, from Spa Grounds
19 The Spa, Scarborough, at noon
20 "Children's Corner," Scarboroug Sands
21 Valley Park, Scarborough
22 Foreshore Road, Scarborough
23 The Keep, Scarborough Castle
24 North Bay, Scarborough
25 Hayburn Wyke
26 Whitby, from the Station
27 Whitby, from Larpool
28 "A Good Catch," Whitby
29 The Piers, Whitby
30 A Shipwreck, Whitby
31 Whitby Abbey, from S. E
32 The Scaur, Whitby
33 Robin Hood's Bay
34 Rigg Mill, near Whitby
35 On the Esk, near Whitby
36 Beggar's Bridge, Glaisdale
37 Runswick
38 Staithes
39 Fishing Boats at Staithes
40 Rievaux Abbey
41 Rievaux Abbey, Nave
42 Bolton Priory
43 Bolton Priory, Nave
44 Bolton Priory, Choir
45 Bolton Hall
46 Lord Fredk. Cavendish s Monument
47 Stepping-stones on the Wharfe
48 The Strid. on the Wharfe
49 On the Wharfe
50 Kirkstall Abbey
51 Kirkstall Nave
52 St. George's Church, Doncaster
53 St. George's Church, Forman Chapel
54 Market Place, Doncaster
55 Conisborough Castle
56 Roche Abbey

ISLE OF WIGHT.

With Reading.

1 Ryde, from the Pier
2 Ryde, the Esplanade
3 Union Street, Ryde
4 Wooton Bridge and Village
5 Wooton Bridge
6 Brading Church
7 Tomb of Little Jane, Brading
8 Little Jane's Cottage, Brading
9 Sandown Bay
10 Shanklin Beach
11 Shanklin Chine—general view
12 Shanklin Chine—the Waterfall
13 Shanklin Chine—Bridge in
14 View across Shanklin Chine
15 Shanklin Village
16 Luccombe Chine
17 Luccombe Chine—the Waterfall
18 Landslip, Bonchurch
19 Bonchurch Old Church
20 Tomb of Shadow of the Cross, Bonchurch
21 Bonchurch Road
22 Bonchurch Pond and Village
23 Ventnor, looking West
24 Ventnor—the Esplanade
25 Ventnor, from Cliff Path
26 Ventnor, "Crab and Lobster Inn"
27 City Missionary Home Ventnor
28 Royal Cottage Hospital, Ventnor
29 Steephill Castle, near Ventnor
30 Steephill Bay, near Ventnor
31 St. Lawrence Church
32 Road at Niton
33 View from Cripple Path, Niton
34 St. Catherine's Lighthouse
35 Blackgang—the Upper Chine
36 Blackgang, from the Beach
37 Freshwater Bay, looking West
38 Freshwater Bay, looking East
39 Arched Rock, Freshwater Bay
40 Cave at Freshwater Bay
41 Watcombe Bay, Freshwater
42 Tennyson's House, Freshwater
43 The Needles, from Scratchell's Bay
44 Scratchell's Bay, Isle of Wight
45 The Needles, Rocks, and Lighthouse
46 Alum Bay, from the West
47 Alum Bay and Pier
48 Yarmouth, from the Bridge
49 Yarmouth—the High Street
50 Newport—general view
51 Newport—the High Street
52 Carisbrook Village
53 Carisbrook Village from the Brook
54 Carisbrook Castle—general view
55 Carisbrook Castle—the Front Towers
56 Carisbrook Castle—King Charles' Window
57 Arreton Church, Isle of Wight
58 Tomb of Dairyman's Daughter, Arreton
59 Cottage of Dairyman's Daughter, Arreton
60 Whippingham Church
61 Osborne House, from the Northwest
62 Osborne House—the Terraces
63 Osborne House, from the Upper Terrace
64 Osborne House, from the Lower Terrace
65 East Cowes and Harbor
66 West Cowes, from the Pier
67 Study of Ferns and Flowers

A TOUR THROUGH CORNWALL.

With Reading.

GERMANY.

With Reading.

8   JAMES W. QUEEN & CO., PHILADELPHIA.

WINDSOR CASTLE.

With Reading.

1 General Ground Plan of the Castle
2 Windsor Castle, from Northwest
3 Gateway of Henry VIII
4 The Horseshoe Cloisters
5 St. George's Chapel, South Front
6 Nave, looking East
7 Choir, looking West
8 Choir Stalls and Royal Pew
9 Interior of Albert Memorial Chapel, looking East
10 Interior of Albert Memorial Chapel, looking West
11 The Round Tower
12 The Norm n Gate
13 Upper Ward, from the

14 Home Park
14 The Guard Chamber
15 The Presence Chamber
16 The Vandyke Room
17 The Rubens Room
18 The Waterloo Chamber
19 The Throne Room
20 The Grand Reception Room
21 St. George's Hall
22 The East Terrace
23 The Corridor
24 The Corridor
25 "Allured to Brighter Worlds and Led the Way."
26 The Dining-Room
27 The Crimson Drawing-Room

28 The Green Drawing-Room
29 The White Drawing-Room
30 The Quadrangle
31 The South Front of Upper Ward
32 The Long Walk, from Snow Hill
33 Snow Hill
34 Virginia Water
35 Ruins at Virginia Water
36 Royal Mausoleum, Frogmore
37 The Interior of the Royal Mausoleum, Frogmore
38 Duchess of Kent's Mausoleum
39 Clewer Parish Church
40 Eton College

A TRIP TO BRIGHTON.

With Reading.

1 Queen's Road and Railway Station
2 West Street and St. Paul's Church
3 The King's Road, looking East
4 The Grand Hotel
5 Regency Square
6 The West Pier
7 Brighton from the West Pier
8 Beach and West Brighton from the Pier
9 Norfolk Hotel and King's Road, Hove
10 Hove New Town Hall
11 Brighton and Esplanade from Band Stand
12 The Beach at the West Pier
13 West Street, from the King's Road

14 New Shelter Hall, Orleans Club-House
15 The Beach and Coastguard Station
16 The Yacht "Skylark" ready to start
17 The Yacht off for a Sail
18 Fish Market on the Beach
19 The King's Road
20 The Old Ship Inn
21 The Aquarium Clock Tower
22 Marine Parade from Aquarium End
23 East Brighton, from the Sea
24 Children on the Aquarium Beach
25 The Chain Pier
26 The Chain Pier from the Pier Head
27 Beach and Sea Wall, from the Pier

28 The Sea Wall, from the Madiera Road
29 Brighton, from the Marine Parade
30 Marine Parade, Kemp Town
31 Aquarium, etc., from the East
32 Brighton Town Hall
33 The Old Steyne Gardens, general view
34 The Victoria Fountain, Old Steyne
35 The Old Steyne, from the Esplanade
36 Royal Pavilion, West Entrance
37 Royal Pavilion, from the East Lawn
38 The Dome Assembly Rooms
39 The Devil's Dyke
40 Swiss Gardens, Shoreham

A TOUR THROUGH NORTH WALES.

With Reading.

1 Llandudno, from the Great Orme's Head
2 Llandudno—Beach and Pier
3 Mostyn Street, Llandudno
4 Conway Castle and Suspension Bridge
5 Banqueting Hall, Conway Castle
6 Llanrwst Bridge
7 Valley at Bettws-y-Coed.
8 Bettws-y-Coed, from the Bridge

9 Fairy Glen, Bettws-y-Coed
10 Conway Falls and Salmon Ladder
11 Stepping-stones, near Bettws-y-Coed
12 Miners' Bridge over the Llugy
13 The Swallow Waterfall.
14 Moel Siabod, from near Capul Curig
15 Snowdon, from Capul Curig Lakes
16 Ogwen Lake

17 Waterfalls of the Ogwen
18 Penrhyn Slate Quarries
19 Bangor—general view
20 Bangor Cathedral—interior
21 Waterfall near Aber
22 Beaumaris Castle—the entrance
23 Menai Suspension Bridge
24 Carnarvon Castle, from the Ferry
25 Carnarvon Castle—interior
26 Llanberis Waterfall
27 Llanberis Church in the Pass

A HOLIDAY GLANCE AT THE SOUTH COAST.

With Reading.

A VISIT TO THE BRITISH MUSEUM.

With Reading.

IRELAND.

DUBLIN, WICKLOW, KILLARNEY, ETC. No. 1.

With Reading.

1 Kingstown Harbor, arrival of mail Steamer
2 An Irish Jaunting Car
3 Sackville Street, Dublin (instantaneous)
4 General Post-Office and Nelson's Pillar
5 Grafton Street (instantaneous), Dublin
6 Bank of Ireland, Old Houses of Parliament, and Statue of Henry Grattan, Dublin
7 Trinity College, Dublin
8 St. Patrick's Cathedral, Dublin
9 St. Patrick's Cathedral, the Choir, Dublin
10 Christ Church Cathedral, Dublin
11 The Four Courts, Dublin
12 The Custom House, Dublin
13 O'Connell's Monument, Glasnevin Cemetery, Dublin
14 The Vice-Regal Lodge, Phœnix Park, Dublin
15 Killiney, and the Vale of Shanganagh, Co. Dublin
16 Bray, and Bray Head, Co. Wicklow
17 The Scalp, Co. Wicklow
18 Cottage in the Dargle (summer), Co. Wicklow
19 Cottage in the Dargle (winter), Co. Wicklow
20 The Dargle, Co Wicklow
21 Enniskarry, Co. Wicklow
22 Powerscourt Waterfall, Co. Wicklow
23 Powerscourt House, Co. Wicklow
24 The Vale of Clara, Co. Wicklow
25 The Valley of Glendalough, and the ruins of the Seven Churches, Co. Wicklow
26 The Vale of Ovoca, Co. Wicklow
27 The Lion Arch, Castle Howard, Vale of Ovoca, Co. Wicklow
28 Johnstown Castle, the Seat of the Earl of Granard
29 Kilkenny, bird's-eye view
30 Kilkenny, the seat of the Marquis of Ormonde
31 Jerpoint Abbey, Co. Kilkenny
32 Lismore Castle, Co. Waterford
33 Ruins on the Rock of Cashel, Co. Tipperary
34 Holycross Abbey, Co. Tipperary
35 Patrick Street, Cork
36 Patrick's Bridge, showing Father Matthew's Statue
37 St. Finn-Barr's Cathedral, Cork
38 Shandon Steeple, Cork
39 The Mardyke Walk, Cork
40 Queenstown Harbor, showing Haulbowline and Spike Islands
41 Sir Walter Raleigh's House, Youghal
42 Blarney Castle, Co. Cork
43 Glengariff Harbor, Bantry Bay
44 Cromwell's Bridge, Glengariff
45 Glengariff Waterfall
46 General View, Lakes of Killarney
47 The Upper Lake, Killarney
48 The Eagle's Nest Mountain, Killarney
49 The Old Wier Bridge, Shooting the Rapids, Killarney
50 The Middle Lake, from Dinnis Island, Killarney
51 The Colleen Bawn Cave, Middle Lake, Killarney
52 Muckross Abbey, Killarney
53 Interior of Muckross Abbey, Killarney
54 Glena Bay, Killarney
55 O'Sullivan's Cascade, Killarney
56 Brickeen Bridge, Killarney
57 The Meeting of the Waters, Killarney
58 Ross Castle, Killarney
59 Derrycunnihy Cottage and Waterfall, Killarney
60 The Gap of Dunloe, Killarney

NORTH AND WEST. No. 2.

With Reading.

1 The Boyne Viaduct at Drogheda, Co. Louth
2 Ancient Cross and Round Tower at Monasterboice, Co. Louth
3 Warrenpoint, Co. Down
4 Rosstrevor Quay and Mourne Hotel, Co. Down
5 Carlingford Lough, Co. Down
6 Armagh, showing Cathedral
7 Donegal Place, Belfast
8 The Albert Memorial, Belfast
9 The Queen's College, Belfast
10 Shane's Castle, Antrim
11 Garron Tower, the seat of the Marchioness of Londonderry
12 The Rope Bridge, Carrick-a-rede, Co Antrim
13 General View of the Great Causeway, Giant's Causeway
14 Lord Antrim's Parlor
15 The Honeycomb
16 The Wishing Chair
17 The Ladies' Fan
18 The Causeway Gate
19 The Giant's Well
20 Dunluce Castle
21 Londonderry, Lough Foyle
22 The Cathedral, Londonderry
23 Walker's Monument, Londonderry
24 Bishop's Gate, Londonderry
25 Horn Head, Donegal
26 Errigal Mountain, Donegal
27 Ruins on Devenish Island, Lough Erne, Co. Fermanagh
28 Holy Well of Tubbernaltha, near Sligo
29 Glencar Waterfall, near Sligo
30 Boyle Abbey, Co. Roscommon
31 Kylemore Castle, the seat of Mitchell Henry, Esq., M. P., Connemara
32 Kylemore Lake, Connemara
33 Ballinahinch and Lake, Connemara
34 The Killaries Bay, Connemara
35 Dugort, Achill, and Slieve Mor Mountains
36 Sunset on Achill Sound
37 Rosserk Abbey, Co. Mayo
38 Cong Abbey, Co. Galway
39 The Fish Market, Galway
40 The Cliffs of Moher, Co. Clare
41 The Spa Well, Lisdoonvarna, Co. Clare
42 The Spectacle Bridge, Lisdoonvarna, Co. Clare
43 Kilkee, Co. Clare
44 The Natural Bridges of Ross, Co. Clare
45 Killaloe, on the Shannon, Co. Limerick
46 Rapids of the Shannon at Castle Connell, Co. Limerick
47 Askeaton Abbey, "the Nave," Co. Limerick
48 Georges' Street, Limerick
49 King John's Castle, and Shomond Bridge, Limerick
50 The Treaty-Stone, Limerick

SCOTLAND.

EDINBURGH.

With Reading.

1 Edinburgh from St. Anthony's Chapel
2 Holyrood Palace, Front
3 Holyrood Palace. Queen Mary's Bedroom
4 Holyrood Chapel
5 Holyrood Chapel, Interior
6 Calton Hill
7 Holyrood Palace and Arthur's Seat
8 Burns' Monument
9 Princes Street from Calton Hill
10 Princes Street and Scott's Monument
11 Statue of Sir Walter Scott
12 East Princes Street from Scott's Monument
13 Old Town from Scott's Monument
14 Bank of Scotland from Scott's Monument

15 Castle and National Galleries from Scott's Monument
16 West Princes Street from Scott's Monument
17 West Princes Street from the Mound
18 Castle Rock from West Princes Street
19 Prince Consort's Memorial
20 St. Mary's Cathedral, Exterior
21 St. Mary's Cathedral, Nave
22 Edinburgh from Craigleith
23 Castle from Grassmarket
24 Castle Esplanade
25 "Mon's Meg." Edinburgh Castle
26 View from Castle looking North
27 Edinburgh from Castle
28 Corner of the West Bow

29 Old Houses in High Street
30 St. Giles' Cathedral, Exterior
31 St. Giles' Cathedral, Interior
32 Market Cross
33 Old Parliament House
34 Royal Infirmary
35 New Medical Class Rooms
36 University Buildings
37 John Knox's House
38 Canongate Tolbooth
39 White Horse Close
40 Queen Mary's Bath House
41 Arthur's Seat
42 Roslin Glen and Castle
43 Roslin Chapel, Exterior
44 Roslin Chapel, Chancel
45 Roslin Chapel, Master Pillar
46 Roslin Chapel, 'Prentice Pillar

THE HIGHLANDS OF SCOTLAND.

With Reading.

Introduction
Glasgow Cathedral
George Square
The University
The Broomielaw
Dunglass Castle
Dunbarton Rock and Castle
Greenock
Inverary
Oban
Staffa, Fingal's Cave
Iona
Glencoe
Ben Nevis
Fall of Foyers
Inverness
Kirkewall (Cathedral)

Duncansby
Dunrobin Castle
Elgin (Cathedral)
Aberdeen
Castle Street and Union Street
King's College
Old Machar Cathedral
Old Brig o' Balgownie
Balmoral
Dark Lochnagar
Dunnottar
Arbroath Abbey
Perth
Dunkeld
Hermitage and Bridge, Dunkle
Pass of Killiecrankie

Blair Athole
Falls, Moness
Taymouth Castle
The Pass of Leny
Callander
Pass of the Trossachs
Loch Katrine
Inversnaid Falls
Loch Lomond
Dunblane
Abbey Craig
Stirling Castle
Dollar (Castle Campbell)
The Devil's Mill
Lochleven Castle
St. Andrews
Dumfermline (Abbey)

THE LOWLANDS OF SCOTLAND.

With Reading.

Introduction
Calton Hill
Holyrood Palace
Scott Monument
Grassmarket
Edinburgh, Old Town
Roslin Glen and Castle
Roslin Chapel
The 'Prentice Pillar
Craigmillar Castle
Tantallon Castle
North Berwick Law
Dirlton Castle
Norham Castle
Twizel Castle
Jedburgh Abbey

The Capon Tree
Kelso Abbey
Floors' Castle
Branxholm Tower
Dryburgh Abbey
Melrose Abbey
Melrose Abbey. East window
Abbotsford
Abbotsford. ' The Study'
St. Ronan's Well
Peebles
Neidpath Castle
Newark Castle
St. Mary's Loch
Gray Mare's Tail
Moffat
Beld Craig, Linn

Caerlaverock Castle
Dundrennan Abbey
Burns' Mausoleum
Lincludon Abbey
On the Nith
Ary (" The Twa Brigs ")
Burns' Cottage
Burns' Monument
The Doon below the Bridge
Stair House
Ballochmyle
Bothwell Castle
Falls of Clyde
Cora Linn
Stonebyres
Linlithgrow

GERMANY.

BERLIN.

With Reading.

1 Railway Station, Alexander Place
2 King Street
3 Place of the Old Palace
4 The Chapel of the Palace
5 The Palace of the Crown Prince
6 Unter den Linden
7 Unter den Linden
8 Royal Palace Guard House
9 Palace of the Emperor
10 Statue of Frederick the Great
11 The Old Museum
12 Front of Old Museum
13 Steps of the Old Museum
14 Old Berlin
15 Old Berlin
16 Oldest house in Berlin
17 The River Spree
18 The Cathedral
19 St. Hedwig's Church
20 The Thiergarten
21 Royal Dramatic Theatre
22 Schiller Monument
23 National Gallery
24 Statue of King William III
25 Statue of Queen Louise
26 A branch of the New Lake
27 The Löwenbrücke
28 Brandenburger Gate
29 Monument of Victory
30 Belle Vue Palace
31 Marble Palace
32 Leipsic Street
33 The French Church
34 Royal Opera House
35 Frederick William University
36 The Arsenal
37 The front of the Arsenal
38 Entrance to Emperor William Street
39 The Exchange
40 The Mint
41 Bruder Street
42 The Fisherbridge
43 Alexander Place Market
44 The Cycle Course
45 Charlottenburg Polytechnic
46 The Military Band

THE RHINE TOUR.

From Cologne to Schaffhausen, with some Peeps at the Black Forest of Germany.

With Reading.

1 Cologne Cathedral, from the Railway Station
2 Cologne Market
3 Bonn, from the Pier
4 Königswinter and the Ferry
5 The Drachenfels, from the opposite bank
6 The Castle of Drachenfels
7 Rolandseck, from the Ferry
8 Remagen, from the West
9 Erpel, from Remagen
10 Andernach Castle
11 The Old Watch Tower, Andernach
12 Coblentz and the Bridge of Boats
13 Ehrenbreitstein Castle
14 Ems and the Bäderlei Hill
15 Ems, looking West
16 Boppard, Old Houses in the Town
17 Boppard, view looking down the Rhine
18 Rheinfels Castle, St. Goar
19 St. Goar, from the opposite bank
20 St. Goarhausen and the Katz Castle
21 St. Goarhausen, from St. Goar
22 The Lurlei Rock
23 Oberwesel, from the Castle Hills
24 Oberwesel, from the Pier
25 Bacharach and St. Werner's Chapel
26 Rheinstein Castle, near Bingen
27 Bingen—the Mouse Tower
28 Mayence Cathedral, from S. W
29 Mayence Cathedral, from the Market Place
30 Mayence and Bridge of Boats
31 Heidelberg Castle, from the Neckar Bridge
32 Heidelberg Castle—the Grand Facade
33 Heidelberg, from the Great Terrace
34 Raft on the Neckar at Heidelberg
35 Baden-Baden, from the Hills
36 The Trinkhalle, Baden-Baden
37 The Conversation House, Baden-Baden
38 Strassbourg Cathedral, etc
39 Hornberg—Street in the Village
40 Hornberg, from the Castle Hill
41 Hornberg—View from the Bridge
42 Niederwasser Village, from the Railway
43 Triberg—General View
44 Triberg—Waterfall of the Failbach
45 Basle, from the Old Bridge
46 The Old Bridge, Basle
47 The Rhinefalls, Neuhausen
48 Neuhausen—the Rhine below the Falls
49 Schaffhausen and the Rapids
50 Schaffhausen, from the Swiss side

SWITZERLAND.

The Northern Lakes and Bernese Oberland.

Tour No. 1, with Reading.

1 Basle, Old Bridge, and "Three Kings" Hotel
2 Basle, The Upper Bridge
3 Basle, The Minster
4 Basle, The Minster, West Door
5 Basle, St. Paul's Gate
6 Neuhausen, The Rhine, above the Falls
7 Neuhausen, The Falls of the Rhine
8 Schaffhausen, from the High Rock
9 Constance, from the Cathedral Tower
10 Constance, The Rhine Bridge
11 Zurich, from the Minster Tower
12 Zurich, View from Flaraten
13 Lucerne, and the Rigi
14 Lucerne, and Pilatus
15 Lucerne, The Hofkirche
16 Lucerne, The "Lion" Monument
17 Lake of Lucerne, View from the Rigi

18 Lake of Lucerne, Tell's Chapel
19 Lake of Lucerne, The Axenstrasse
20 Brunnen
21 The Sarnen Sea
22 Handeck, the Châlet
23 Handeck, Falls of the Aar
24 Grimsel Lake and Hospice
25 Rosenlaui, Falls of the Reichenbach, No. 1
26 Rosenlaui, Falls of the Reichenbach, No. 2
27 Rosenlaui, The Wellhorn and the Wetterhorn
28 Interlaken, View at the Rugen
29 Interlaken, Hotel Beau-Rivage
30 Interlaken, The Giessbach
31 Lauterbrunnen, The Staubbach
32 View at Mürren
33 Wengern Alp and Little Scheideck Pass
34 Grindelwald, The Wetterhorn

35 Grindelwald, The Eiger, etc
36 Grindelwald, The Upper Glacier
37 Grindelwald, on the Eismeer
38 Spietz, Lake of Thun
39 Kandersteg, Bear Hotel and Mt. Lohner
40 Gasterin Thal, The Schildhorn
41 Blumlis Alp, from path to the Lake
42 Thun, from Belle Vue Pavilion
43 Thun, from the Churchyard
44 Thun, The Castle, etc
45 Fribourg, Lower Town Fountain
46 Fribourg, The Suspension Bridge
47 Fribourg, The Cathedral
48 Berne, from Rosengarten
49 Berne, from Schaenzli
50 Berne, The Clock Tower

List of Slides to Illustrate the Reading.

No. 2.

1 Geneva. From Bridge over Rhone
2 Chamounix and Mount Brevent
3 Tête Noire. First Peep of Mont Blanc
4 Tête Noire. Salvan Route
5 Tête Noire. From Roche Percée
6 Tête Noire Valley
7 Mer de Glace. From the Flegere
8 Mer de Glace
9 Mer de Glace
10 Dome de Goutez. Glacier des Bossons
11 Mont Blanc. From Glaciers
12 Mer de Glace
13 Vernayaz. Pissevache Cascade
14 Vernayaz. Gorge du Trient
15 Vernayaz. Gorge du Trient
16 Zermatt and the Matterhorn
17 Zermatt. The Riffelhaus

18 Zermatt and the Matterhorn
19 The Matterhorn. From Mettelhorn
20 The Matterhorn. From Gornergrat
21 The Lyskamm and Twins
22 Monte Rosa
23 Brieg, Simplon Gorge, and Mont Leone
24 Brieg and Bel Alp
25 St. Gothard. Pont du Diable
26 Pont du Diable
27 Pont du Diable. St. Gothard
28 Hospenthal and Mont Tibia
29 Amstag
30 Maderaner Thal
31 Maderaner Hüfihorn, and Breithorn
32 Maderaner. Stauerbach Cascade
33 Amstag. From the Reuss Bridge
34 Viesch Glacier and Finsteraarhorn
35 Aletsch Horn, Jungfrau, Little Aletsch Glacier

36 Meerielensee, Aletsch Horn and Glacier
37 Furca, Todtensee, and Finsteraarhorn
38 Rhone Glacier and Hotel
39 Rhone Crevasse and Glacier
40 Grindelwald. Ice Cave
41 Kandersteg, Blumlis Alp and Oexhinen Lake
42 Loëche les Bains and Gemmi Pass
43 Loëche les Bains. Ladder Pass
44 Sion. Rhone Valley
45 Vevay, Montreux, and Dent du Midi
46 The Castle of Chillon. Front Entrance
47 The Castle of Chillon. From the Water
48 Ouchy Hotel. Beau Rivage
49 Lausanne. From the Promenade
50 Lausanne. Castle and Cathedral

THE ENGADINE, SWITZERLAND.

With Reading.

1 Street in Coire
2 Coire Cathedral—Interior
3 Thusis—General View
4 In the Schyn Pass
5 The Pont du Solis, Schyn Pass
6 Tiefenkasten Village
7 Molins Village
8 Silvaplana Village and Lake
9 Silvaplana, from the Meadows
10 Haymaking going on
11 Group of Haymakers and Ox Wagon
12 The Surlei Cascade
13 Footpath along Lake Silvaplana
14 Lake Silvaplana, from the Woods
15 Sils Maria Village
16 Sils Maria, from the South
17 An Ox Load of Hay
18 Lake of Sils and Isola Village
19 Maloja, from Lake of Sils (Cattle)
20 Maloja, from Promenade des Artistes
21 Oxen Carting Home the Hay
22 Campfer—the Lakes and Piz Margua
23 Campfer—the Valley and Lake
24 The River Inn, near St. Moritz Bad
25 St. Moritz, from Johansberg
26 St. Moritz Village, from the South
27 St. Moritz Bad and the Lake
28 Falls of the Inn, St. Moritz
29 The Lower Valley, from St. Moritz Hill
30 Path through the Woods to Pontresina
31 Young Engadine Cattle
32 Pontresina and Piz Languard, from West
33 Road through Pontresina Village
34 Washerwoman at a Street Fountain
35 Pontresina and the Roseg Valley
36 The Roseg Glacier
37 In the Pine Woods, Pontresina
38 Pont Ota, Pontresina
39 The Morterasch Glacier
40 Samaden Village
41 A Diligence
42 Bevers Village
43 Ornamented Houses in Bevers Village
44 Ponte Village, from the Meadows
45 Madulein and the Inn, from Ponte
46 The Albula Hospice or Inn
47 The Weissenstein Valley, Albula Pass
48 Bergun Village, Albula Route
49 Street in Filisur Village
50 Coire—General View

THE FOREST CANTONS OF SWITZERLAND.

Embracing a Trip over the Rigi by the Cog-wheel Railway, Pictures of the Beautiful Lake Lucerne, and the Roads up the St. Gothard.

With Reading.

1 Lucerne—the Lake Steamers, etc.
2 Lucerne and Mount Pilatus
3 Lucerne—The Schweizerhof Hotel
4 Lucerne—The Kapell Bridge and Old Tower
5 Lucerne—The Lion Monument
6 The Glacier Garden, Lucerne
7 The Swan Gardens on the Lake Lucerne
8 Lucerne and the Rigi, from the Gütsch
9 Stanzstad and the Bürgenstock
10 Stanzstad—The Old Watch-Tower
11 Valley of Lungern, from the Brünig Pass
12 Kussnacht, Lake Lucerne
13 Weggis and the Rigi
14 Vitznau—The Mountain Locomotive, etc.
15 Vitznau, from the Railway
16 The Schnur-Tobel Bridge, with Train
17 The Grubis-Bahm Waterfalls Rigi
18 View from the Railway at Romiti Felsenthor
19 Rigi Karltbad—The Hotels, etc.
20 Chapel at the Sisters Fountain, Karltbad
21 Rigi Staffel
22 Rigi Culm, from Rigi Staffel
23 Klosterli Village on the Rigi
24 Bridge over the Rothenfluhbach Ravine
25 Arth Rigi Railway—Cutting through Forest
26 Lake Lowerz and the Mytons
27 Valley of Goldau and the Rossberg
28 Goldau Church
29 Arth, foot of the Rossberg
30 Schwyz Market Place and the Mytons
31 Brunnen, from the Axenstrasse
32 The Axenfels and St. Gothard Line
33 Gersau, from the Pier
34 Gersau from the Hills
35 Lake Lucerne, from Tell's Platte
36 Tell's Chapel, Lake Lucerne
37 The Axenstrasse, above Tell's Chapel
38 The Axenstrasse Tunnel, near Fluelen
39 Fluelen, from the Railway Bridge
40 Altdorf—Statue of William Tell
41 Amsteg and the Burgenstock Mountain
42 Amsteg, from the Bridge
43 Wasen—General View
44 View from Wasen Church yard
45 Rolirbach Falls and Bridge, near Wasen
46 Gescheneu, looking up the St. Gothard Pass
47 Geschenen, from the Railway Bridge
48 St. Gothard Pass—the Schöllinen Gorge
49 The Devil's Bridge, St. Gothard Pass
50 Hospenthal and the River Reuss

ITALY.

With Reading.

1 Turin. Palazzo Carignano
2 Turin. Church of Gran Madre di Dio
3 Turin. Capuchin Mount and Monastery
4 Milan Cathedral
5 Orta Lake
6 Baveno. Lake Maggiore
7 Varenna. Lake Como
8 Venice. Cathedral of St. Mark's, Facade
9 Venice. Reliefs in Porphyry of Knights
10 Venice. Palace of the Doges
11 Venice. Palace of the Doges, Bronze Fountain in Courtyard
12 Venice. The Bridge of Sighs
13 Venice. Grand Canal
14 Venice. Ponte Rialto
15 Venice. Ca d'Oro, Golden Palace
16 Bologna Church of San Giacomo Maggiore
17 Florence and River Arno
18 Florence. Cathedral and Campanile from Palazzo Vecchio
19 Florence. Palazzo Vecchio
20 Florence. Uffizi and Palazzo Vecchio
21 Florence. Loggia dei Lanzi
22 Florence. Church of Santa Croce
23 Florence. Triumphal Arch at Porto Gallo
24 Arezzo. Birthplace of Petrarch
25 Naples, from S. Elmo
26 Naples. Harbor from the Arsenal
27 Naples. S. Elmo and Marina
28 Naples. S. Lucia and Castello dell'Ovo
29 Naples. Piazza del Plebiscito
30 Naples. Palazzo Reale
31 Naples. Palazzo Reale—the Scala
32 Naples. Church of S. Francesco di Paolo
33 Naples. Villa Nazionale
34 Naples. Monument in the Piazza de Martiri
35 Pompeii. View with Vesuvius
36 Pompeii. The Forum
37 Pompeii. The Basilica
38 Pompeii. The Temple of Venus
39 Pompeii. The Pantheon
40 Pompeii. The House of the Small Fountain
41 Pompeii. The House of the Faun
42 Pompeii. The Amphitheatre
43 Pompeii. The Street of the Tombs
44 Palermo Cathedral
45 Pisa, Baptistery, Cathedral, and Leaning Tower
46 Pisa. Leaning Tower
47 Lucca Cathedral
48 Genoa. Above the Railway Station
49 Genoa. Palazzo Ducale
50 Genoa. Christopher Columbus

THE ITALIAN LAKES.

An Entirely New Series of Fifty Views.

With Reading.

1 Street in Bellinzona
2 Locarno from the Pier
3 Locarno. the Grand Piazza
4 Madonna del Sasso from the Town
5 Madonna del Sasso from the Heights
6 Cannobo from the Steamer
7 Maccagno from the Lake
8 Luino from the Lake
9 Intra from the Pier
10 Laveno from the Lake
11 Pallanza, Piazza Garibaldi
12 Villa Clara, Baveno
13 Baveno from the Hills
14 Stresa, and Lake Maggiore from the Hills
15 Isola Bella from the Shore
16 Isola Bella, the Grotto in Gardens
17 Arona from the Lake
18 Lake Orta from the South
19 The Lake at Orta
20 Market Place, Orta
21 Street in Orta
22 Island of San Guilio, Lake Orta
23 Madonna del Monte, near Varese
24 Lugano and Monte Bré
25 Lugano and Monte San Salvatore
26 Lugano and St. Lorenzo Church
27 Oxen Cart, Lugano
28 Lugano from the Lake
29 The Wood Market, Lugano
30 Fresco by Luine at the Lugano
31 Osteua, Entrance to the Grotto
32 Poreezza, the Landing-Place, etc
33 Menaggio, the Town, and Lake Como
34 Bellaggio, the Via Serbelloni
35 Bellaggio, general view
36 View from Park of Villa Serbelloni
37 Tropical Plants in Garden of Villa Serbelloni
38 Lake Como, Lavedo and Island Comacina
39 View from above Sala, looking north
40 Torrigia from the Lake
41 Como, the Harbor and Landing-Place
42 General view of Como
43 Como and the Lake from Baradello Hill
44 Milan Cathedral from the Piazza
45 View on the roof of Milan Cathedral
46 Lecco, general view
47 Lake Lecco and Mountains from the Bridge
48 Lecco from the Shore
49 Gravedona, from the Lake
50 Colico, the Diligence Office, etc.

THE RIVIERA.

From Marseilles to Genoa.

With Reading.

BELGIUM.

ROME.

With Reading. By Dr. Edgar.

1 View from the French Academy
2 St. Peter's from the Vatican Gallery
3 St. Peter's
4 St. Peter's Facade and Dome
5 St. Peter's, The Interior
6 The Vatican
7 Chiaramonti Corridor, The Vatican
8 The Vatican Ariadne
9 Ruins on the Palatine Hill
10 Temple of Vesta
11 Temple of Castor and Pollux
12 San Lorenzo, outside the Walls
13 Temple of Faustina from Palatine Hill
14 View from the Palatine Hill
15 Temple of Saturn

16 Arch of Constantine
17 Arch of Titus
18 Arch of Titus, Bas-relief (7 Candlesticks)
19 Arch of Titus, Bas-relief (the Chariots)
20 Porta San Maggiore
21 Porta San Lorenzo
22 Porta San Paolo
23 Porta San Giovanni
24 Tomb of Cecelia Metella
25 Basilica of Constantine (distant)
26 Basilica of Constantine (near)
27 Island in the Tiber
28 Palazzo Quirinale
29 Palazzo del Laterano
30 Villa Medici
31 St. Angelo
32 Fontana Paolino

33 Fontana di Trevi
34 The Coliseum
35 The Coliseum, the Interior
36 Piazza Navona
37 Piazza Colonna
38 Steps of Piazza di Spagna
39 View from Monte Pincio
40 Piazza del Popolo
41 Column of Immaculate Conception
42 Pincio Gardens
43 Pincio Gardens, Fountain of Moses
44 The Pantheon
45 San Paolo
46 Forum Romanum
47 Forum of Trajan
48 Column of Phocas
49 The Capitol
50 Santa Maria Maggiore

THE CITY OF FLORENCE.

With Reading.

1 Church of St. Maria, Novella
2 Pont Trinita
3 The Vecchio Bridge
4 View from the Vecchio Bridge
5 The Vecchio Palace and Logia
6 Court of the Vecchio Palace
7 Great Hall of the Vecchio Palace
8 The Cathedral, from Vecchio Tower
9 Fountain and Piazza della Signoria
10 Logia de Lanzi
11 Florence Cathedral—the Front

12 Giotto's Tower, etc., from the East
13 The Piazza del Duomo
14 The Baptistry—Florence
15 Front doorway of Florence Cathedral
16 Florence, from San Miniato
17 Monastery and Cemetery of San Miniato
18 View from Piazza, Michael Angelo
19 Florence and the Arno, from Piazza
20 Uffizi Colonnade and Vecchio Palace
21 East Corridor of the Uffizi Gallery
22 Statue of Venus, Tribune Gallery

23 View from the Vecchio Gallery
24 Pitti Palace, from Boboli Gardens
25 Florence, from the Boboli Gardens
26 A Statue Group in the Boboli Gardens
27 The Pitti Palace—Exterior
28 Church of St. Croce
29 Court of Bargello Museum
30 The Strozzi Palace
31 Church of St. Marco
32 Arch of Gallio
33 The Villa Palmeria
34 Distant View from Fiesole
35 Carthusian Monastery and Monks
36 The Cascine, near Forence

THE MEDITERRANEAN.

With Reading.

1 Gibraltar
2 Gibraltar, from Europa Point
3 Gibraltar Town and Bay
4 Barcelona. The Harbor
5 Marseilles from Notre Dame de la Garde
6 Marseilles. Cathedral of Notre Dame de la Garde
7 Marseilles. Fort Napoleon, etc
8 Marseilles. View in the Harbor
9 Marseilles. Museum Fountain

10 Cannes, from La Californie
11 Cannes, from Mount Chevalier
12 Cannes. Mount Chevalier from the Beach
13 Cannes. Cathedral Tower, Mount Chevalier
14 Antibes
15 Nice, from Ville Franche Road
16 Nice. Jardin Anglais
17 Corsica
18 Nice. The Bay
19 Nice. Promenade des Anglais

20 Nice. View in the Harbor
21 Nice. View in the Harbor with Piers
22 Nice. Les Quais
23 Monaco. Monte Carlo
24 Monaco. Monte Carlo Gardens
25 Monaco. Monte Carlo Gardens
26 Mentone. Old Town from Harbor
27 Mentone. Promenade
28 Genoa, from above the Railway Station

29 Genoa. Christopher Columbus
30 Naples, from San Elmo
31 Naples. Bay and Vesuvius
32 Naples. Marina and San Elmo
33 Sorrento, from Capodimonte
34 Capri. The Marina
35 Capri
36 Amalfi
37 Messina, from the Hills

38 Malta. Grand Harbor
39 Malta. Marsa Muscat
40 Malta. Valetta
41 Malta. Valetta. Church of St. John
42 Malta. Entrance to Grand Harbor
43 Malta. H. M.'s Fleet and Troop-ship
44 Algiers. General View from Harbor

45 Algiers with Boulevards
46 Algiers, from Marengo Gardens
47 Algiers. Palms in Jardin d'Assay
48 Algiers. Palms in Jardin d'Assay
49 Algiers. Interior of Arabesque House
50 Algiers. Gorge of Scheffa

ANCIENT GREECE.

Lecture by P. H. Farez.

1 Map of Greece
2 Mycenæ and Palace of Agamemnon
3 A Centaur
4 Ruins on the Plains of Troy
5 View of the Shores of Greece
6 The Piræus, Athens in the distance
6½ Course of the Ulyssus
7 Grotto of Pantili
8 Panorama of Ancient Athens
8½ Areopagus or Mars Hill
9 Tribune of Demosthenes
10 The Acropolis
11 Parthenon
12 Erechtheion
13 Pandrosium

14 Temple of Theseus
15 Temple of Wingless Victory
16 Temple of Jupiter
17 Tower of the Winds
18 Theatre of Bacchus
19 Gymnasium or School
20 Temple of Neptune at Corinth
21 Lantern of Diogenes
22 Colossus of Rhodes
23 Interior of Arcade at Delphi
24 Costume of Greeks
25 Costume of Greek Ladies
26 Shepherds and dresses
27 Agriculture—Reaping
28 Sports of Children
29 Marriage Procession
30 Panathenaic Procession

31 A Banquet
32 Boxing
33 Horse-racing
34 Throwing the Discus
35 Greek Warriors
36 Chariot and Warriors
37 Macedonian Phalanx
38 Battering Ram
39 Catapult
40 Storming a City
41 Demosthenes
42 Statue of Jupiter Olympus
43 Charon
44 The Nine Muses
45 Perseus with Medusa's Head
46 Laocöon
47 Portrait of Homer

SPAIN.

With Reading.

1 Gibraltar, Our Courier
2 Gibraltar, How we saw Gibraltar
3 Gibraltar, The Bay
4 Gibraltar, Alameda
5 Gibraltar, Catalan Bay
6 Gibraltar, Bridge of Thunder
7 Cadiz, The Cathedral
8 Seville, The Cathedral and City
9 Seville, The Alcazar Gardens
10 Seville, Hall of Ambassadors
11 Seville, Court of Sultanas
12 Seville, The Bull Ring
13 Seville, A Bull Fight
14 Seville, A Bull Fight
15 Cordova, The Town
16 Cordova, Court of Oranges

17 Cordova, Interior of Mosque
18 Cordova, Trionfo Monument
19 Toledo, with the Alcazar
20 Toledo, from the N. W
21 Madrid, Royal Palace
22 Madrid, National Museum
23 Madrid, Fountain of Alcala
24 Madrid, The Escurial
25 Valladolid, an Antique Street
26 Burgos, from the River
27 St. Sebastian
28 Saragossa, The Market
29 Lexida
30 Manresa, The Old Town
31 Barcelona, The Harbor
32 Tarragona, The Cathedral
33 Valencia
34 Granada, Elms in Alhambra Grounds
35 Granada, Cielo Bajo

36 Granada, The Alhambra, from San Nicolas
37 Granada, Court of Lions, Moorish Palace
38 Granada, Hall of Two Sisters
39 Granada, Hall of Justice
40 Granada, Moor's Seat
41 Granada, Sierra Nevada from Adarbes
42 Granada, Tower of Peaks
43 Granada, Water Tower
44 Granada, Gipsy Prince
45 Granada, Gipsy Girl
46 Granada, Group of Gipsies
47 Loja
48 Malaga, Cathedral and Harbor
49 Malaga, The Covered Market
50 Malaga, The Harbor

EGYPT.

CLEOPATRA'S NEEDLE;

Or, the History of Egypt as told by its Monuments.

With Reading. By H. Gore, C. E.

HOLY LAND, OR PALESTINE.

A SET OF SIXTY PLAIN PHOTOGRAPHIC VIEWS AND LECTURES.

INDIA.

With Reading.

ATHENS AND THE PIRÆUS.

FROM LONDON TO MONT BLANC.

With Lecture.

JAPAN;

or, a Visit to the Land of the Rising Sun.

With Reading.

A VISIT TO HOLLAND.

Illustrated by a series of Fifty Photographs.

WANDERINGS IN BIBLE LANDS.

With Reading.

THE SIGHTS OF ROME.

With Reading.

THE BAY OF NAPLES.

With Reading.

POMPEII, PAST AND PRESENT.

Including a few Pictures from drawings, showing the Houses and Temples as they appeared before the destruction of the City.

With Reading.

1 Vesuvius, etc., from Castellammare
2 General view of Pompeii—restored
3 The Basilica, or Hall of Justice
4 The Temple of Apollo
5 The Forum from the South
6 North end of the Forum—restored
7 Columns of the Temple of Jupiter
8 South end of the Forum—restored
9 Street of Abundance
10 A Street Fountain—restored
11 House of Holconius
12 Collection of Sculptures and Ornaments
13 View of the Large Theatre
14 The Small Theatre, from the stage
15 Temple of Isis
16 Temple of Æsculapius

17 Court of the Gladiators
18 Excavators clearing the ruins
19 House of Cornelius Rufo
20 General View of Pompeii
21 The Ampitheatre, general view
22 The Stabian Street, and Stepping Stones
23 A Baker's Oven and Mills
24 House of Marcus Lucretius
25 The Street of Nola
26 A Wine Shop in the Street of Nola
27 A Wine and Provision Shop—restored
28 A Bread Bakery, Street of Nola
29 House of the Chase
30 House of the Fawn
31 House with Columns—restored
32 The Street of Fortune
33 Temple of Fortune—restored

34 A Room in the Baths of the Forum
35 House of the Poet
36 The Poet's House—restored
37 Frescoes inside of a room
38 A Mosaic and shell fountain
39 The House of Pansa
40 Interior of the House of Pansa—restored
41 Statue of Diana
42 House of Sallust
43 Bronze Ornaments d Money Chests
44 Gate of Herculaneum
45 Gate of Herculaneum—restored
46 Seat of Mamia, outside the gate
47 The Street of Tombs
48 Villa of Diomede
49 A Portico and Inner Garden—restored
50 The Museum at Pompeii

THE GREAT PYRAMID.

(From direct negatives.) By Prof. PIAZZI SMYTH.

1 New Excavations of King Shafre's Granite Tomb.
2 Second Pyramid, from the Libyan Desert.
3 Second and Third Pyramids.
4 Entrance Passage of King Shafre's Tomb.
5 Alee Dobre cogitating in King Shafre's Tomb.
6 Well Chamber of King Shafre's Granite Tomb at 4 min. to 12.
7 Well Chamber of King Shafre's Granite Tomb at 12.
8 Well Chamber of King Shafre's Granite Tomb at 12.4.
9 The Western Aisle of King Shafre's Granite Tomb.
10 The Great Sphinx.
11 Coffer in King's Chamber of Great Pyramid.
12 Coffer in King's Chamber of Great Pyramid, and Ghosts of Arabs.
13 The Broken South-East Corner of Coffer in King's Chamber.
14 Base of Niche in the Queen's Chamber, Great Pyramid.
15 Mouth of Entrance Passage leading into Great Pyramid.
16 The Third and Fifth Pyramids of Jeezeh.
17 North Front of Great Pyramid.
18 Mouth of the Entrance Passage, Side View of.
19 The Angle Stones over the Mouth of the Entrance Passage.
20 Ibrahim, the Cook, at the door of his Tomb Kitchen.
21 Sand Slope leading to West Entrance into King Shafre's Granite Tomb.
22 Side View of Beginning of Slope, Entrance Passage
23 Distant View of the Great Pyramid and the Second Pyramid.
24 A Portion of the Granite Casing in situ of the Third Pyramid.
25 A Burial Cave in the East Side of the Great Pyramid Hill.

26 Alee Dobre, Pyramid Arab and Guide, at East Tombs.
27 Vertical Section of Great Pyramid, as engraved for Rl. Obserzatory, Edinburgh.
28 The Great Pyramid and the Remains of the Old Causeway thereto.
29 The North-East Corner of the Great Pyramid.
30 The Palm Trees of Egypt.
31 The Eastern and Northern Faces of the Great Pyramid.
32 The Northern or Entrance Face of the Great Pyramid.
33 The Great Pyramid and the Second Pyramid from the North.
34 Alee Dobre, Pyramid Arab, at East Tombs, Pyramid Hill.
35 The Close of the day at the Pyramid Hill.
36 The Day Guard at East Tombs.
37 The Second Pyramid from King Shafre's Granite Tomb.
38 All the Pyramids of Jeezeh, from the South.
39 The Southern Hill, and Three-Tree Valley.
40 The Corner-Stone Socket, North-East Corner.
41 The South-East Corner Socket—Hole of the Great Pyramid.
42 South-West Socket—Hole of the Great Pyramid.
43 The North-West Socket—Hole of the Great Pyramid.
44 The North-East Socket, repeated, but with measuring rods.
45 The Great Pyramid and its Hill of Rifled Tombs.
46 Part of the Western Excavated Enclosure of the Second Pyramid.
47 Abdul Samud, Sheik of the Northern Pyramid Village.
48 Engraved Vertical Section of King's Chamber and Howard Vyse's Chambers.

PICTURESQUE SCENERY OF DEVONSHIRE.

50 Views.

Barnstable, the Old Bridge.
Lynmouth, from the Footpath.
"   Old Cottages from the Pier.
Woodside Cottage and Bridge, River Lyn.
The Falls at Watersmeet, River Lyn.
Lyn Cliff and Lynmouth.
Lynton, the Village.
Valley of Rocks, Lynton.
Castle Rock, near Lynton.
Ilfracombe from Hillsborough.
"   "   Capstan Hill.
Beach at Ilfracombe.
Bideford from Fort Hill.
Westward Ho! from the Pebble Bridge.
Clovelly from the Pier.
Street in Clovelly, looking up.
"   "   "   down.
Cliffs and Beach at Clovelly.
Torrington, from the Castle Hill.
Okehampton, General View.
Chagford from the Hills.
Old Water Mill at Chagford.
Vixen Tor, Dartmoor.
Tavistock, General View.
"   the Abbey Buildings.

Lydford Waterfall.
Plymouth, View from the Hoe.
Plymouth Hoe from the Pier.
Fishing Boat Sailing out of Harbor
Plymouth, the Guildhall.
Ivybridge, View on the River Erme.
Totnes, the High Street.
Berry, Pomeroy Castle.
"   "   "   (Interior).
Dartmouth and the Harbor.
View up the Dart from Dartmouth.
Kingswear from Dartmouth.
Brixham, the Harbor, etc.
Brixham Trawlers.
Berry Head from South Port.
Torquay and the Harbor.
"   from the Warren.
Natural Arch, Torquay.
Anstis Cove, near Torquay.
Babbacombe Bay and Beach.
Teignmouth, General View.
Dawlish from the West Cliff.
Exmouth, the Strand.
Exeter Cathedral, the West Front.
"   "   the Interior.

THE CHANNEL ISLANDS.

JERSEY, GUERNSEY, SARK, AND ALDERNEY

A Lecture Series of 60 Pictures.

The Steamer entering Guernsey Harbor.
St. Helier's Harbor, Jersey.
Jersey Harbor and Town, from Pier Road.
St. Elizabeth Castle, from the Shore.
Royal Square, St. Helier's.
Broad Street,   "
St. Aubin's, general view.
"   Bay, from St. Aubin's.
The Harbor, St. Aubin's.
A Jersey Farm House.
Portelet Bay and Janverin's Tower.
St. Brelade's Bay.
"   Church and Fishermen's Chapel.
Corbiere Rocks and Lighthouse.
St. Peter's Valley, the Valley Hotel.
Piemont, from the Sands.
Grave de Lecq, from the Heights.
"   "   Exterior of Caves.
A Jersey Cow.
Bon Nuit Bay, Jersey.

Lane and Cottage at Bouley Bay.
Rozel Harbor, Jersey.
Cottage at Rozel Bay.
Road at Rozel Manor.
Anne Port, Jersey.
Druid's Temple, Gorey.
Mount Orguel Castle.
"   "   "   from the Village.
"   "   "   from Gorey Pier.
Gorey, the Bay, etc., from the Castle.
Mail Boat leaving Jersey.
Arrival at Guernsey—Scene on landing
St. Peter's Port, from the Harbor.
The Town Church, from the Old Harbor
The Constitution Steps.
St. Peter's Port, from the Cliffs.
The Harbor at Guernsey.
Jerbourg Point, from Petite Point.
Water Lane, St. Martin's.
Moulin Huet Bay.
Moulin Huet Bay, the Rocks and Beach.
Petit Bot Valley.

Le Moye Point, from the Gouffre.
Le Moye Harbor.
A Primrose Bank
Boadeaux Harbor, St. Sampson's.
Granite Quarries near St. Sampson's.
Creux Harbor, Sark.
Road from the Harbor, Sark.
The Coupé, Sark.
Gouliot Rocks and Caves.
A Sark Well.

Isle of Marchands, or Brechou.
Moie de Moulin and the Autelets.
Natural Arch, Moie de Moulin.
The Creux Terrible
Rocks in Dixcart Bay.
Sark Cottages
Alderney, Fort Torgee and Crabbe Bay.
The Alderney Cow at Home—Fort Gros-
nez and Pier in the distance.

A New Series of 40 Photographs, Illustrating
A VISIT TO WESTERN NORWAY.

Nordfjord, Oldendal, Brynestad Sæter.
" View down Oldendal.
" Foot of Bricksdal Glacier.
" Children and Kids, Bricksdal.
" View up the Loen-Vand.
" Icefall, Kjendalsbrae, Lodal.
" on the Loen-Vand.
Geiranger Fjord, the Knivslaafosse.
Waterfall on the Geiranger Fjord.
View up the Geiranger Fjord.
Söndmore, Near Fibelstad-Hougen.
" Fibelstad-Hougen.
" Pass to Oie, and the Oienibba.
" Oie and Norangsdal.
" On Pass, Orstenvik to Standal.
" Standal, and the Hjorendfjord.
Molde and Moldefjord, from the Ræk-
næshaug.
Molde, from one of the Islands.
Molde and Moldefjord from the Varde.
Romsdal, Hotel Aak and the Romsdal-
shorn.

Romsdal, the Trolltinder.
" from top of Middags-Hougen.
" View on the Rauma.
" near Horgheim.
" the Vermofos.
Jotunheim, the Semmeltind.
" Gjendebod and Svartdalspig.
" Group at Gjendebod.
" Gjendebod from Svartdal.
" Eldsbugaden.
" from the Skinegg looking W.
Sognefjord, the Vettisfos, from below.
" the Afdalfos, near Vetti.
" the Gjellefos, near Vetti.
" from the Hotel Door, Gub-
vangen.
Hardangerfjord, Odde and Sor Fjord.
" Married Woman, Odde.
" Girl, Odde.
" Skjæggedalsfos.
" "

ROUND THE WORLD WITH A CAMERA.
60 Views.

Chart.
London.
Gibraltar
Naples.
Valetta.
Constantinople.
Port Said.
Cairo.
Pyramid and Sphinx.
Group on Board the "Cuzco."
Diego Garcia.
Group on Diego.
New Plymouth, New Zealand.
Wharè.
Group.

Bush.
"Chapman's" (Bush and River).
Bush.
Maori Girls.
Auckland Harbor.
Tauranga.
White Terrace (A.)
" " (B.)
" " (C.)
" " (D.)
" " (E.) Mud Hills.
Pink " (F.)
" " (G.)
" " (H.)
Tiki-teri.

White Island.
Group of Maories.
"Sugar Loaves," New Plymouth.
Sea Piece.
Parihaka, Maori Capital.
Wellington.
Auckland, from North Shore.
Waiwera.
Trees at Honolulu.
Hotel at "
View from Tower of Hotel.
Palace, Honolulu.
San Francisco.
At Clarke's, California.
Grizzly Giant.

Wawona—Big Tree.
Mist in the Yosemite.
From Photographer's Point.
Merced River.
Mirror Lake.
North Dome and River Merced.
Horseshoe Falls, Niagara.
American Fall, "
Rapids, "
Broadway, New York.
Brooklyn Bridge.
Washington—the Capitol.
Iceberg.
Mersey.
Home.

GABRIEL GRUBB.

Colored. The set $18.00.
Uncolored. The set $11.00.

The Story of the Sexton who was Stolen by the Goblins With Special Reading, published by Chapman & Hall.

Introduction.
Panorama—An old Abbey Town.
He sat himself down on a flat tombstone.
Close to him was a strange, unearthly figure.
Playing at leapfrog with the tombstones.
He found himself in a large, dark cavern.
A thick cloud rolled gradually away.
A crowd of little children were gathered round.
He was wet and weary.
Then he sat down to his meal.

The fairest and youngest child lay dying.
The father and mother were old and helpless now.
The few who yet survived then knelt by their tomb.
A rich and beautiful landscape was disclosed.
Lying at full length on the tombstone.
The lantern, the spade, and the wicker bottle.
He told his story to the clergyman and to the Mayor.

JANE CONQUEST.

Finely colored. The set $15.00.
Uncolored. The set $9.50.

This poem has been re-written by Dr. Croft, late Honorary Managing Director of the Royal Polytechnic.

And her child was dying.
Up to her feet rose she.
She saw a gallant ship.
She sank to her knees and made.
Angel effect. Take though my boy.
The snow lay deep.
Stood the old gray church.
And grasped the rope, sole cord of hope.

And then it ceased its ringing.
Midst the breakers.
Saved from the wreck.
Within the silent, darkened room.
Sinks fainting on the ground.
He finds her lying there.
'Tis Harry Conquest.
The suffering boy, her darling boy.

No. 5, angel effect, repeated with No. 16.

NEW YORK CITY.

No.
1 Ocean Steamer.
2 ⎫ New York, looking South.
3 ⎪ New York, looking North.
4 ⎪ New York, looking East.
5 ⎭ New York, looking West.
6 Barge House.
7 Castle Garden.
8 Produce Exchange.
9 Mills Building.
10 Wall Street.
11 Treasury, Washington Statue.
12 The Stock Exchange.
13 Post-Office.
14 Broadway, from Post-Office.
15 City Hall.
16 Court House.
17 The Tombs.
18 Elevated Railway and Cooper Institute.
19 New York University.
20 Public Schools.
21 Broadway, from Stewart's.
22 Fifth Avenue Hotel.
23 Fifth Avenue.
24 Florence Flats.
25 Stewart's House.

No.
26 Union League Club.
27 Jewish Synagogue.
28 St. Patrick's Cathedral.
29 Vanderbilt Mansion.
30 Columbia College.
31 The Mall, Central Park.
32 The Lake, Central Park.
33 The Terrace, Central Park.
34 The Egyptian Obelisk Central Park
35 Highbridge Aqueduct.
36 Elevated Railroad.
37 Oyster Market.
38 West Bridge.
39 Dust Barge.
40 Canal Barges.
41 River Steamer.
42 River Steamer, Interior
43 Brooklyn Bridge.
44 Brooklyn Bridge Footway
45 Fulton Ferry Boat.
46 Fulton Ferry House.
47 Brooklyn Court House and City Hall
48 Greenwood Cemetery, Brooklyn.
49 Morse's Monument.
50 Soldier's Monument.

A WEEK IN VENICE.

1 St. Mark's Tower (The Campanile).
2 Venice from St. Mark's Tower.
3 The Clock Tower, Grand Piazza.
4 St. Mark's Cathedral, from the Piazza.
5 St. Mark's Cathedral, Northwest Corner.
6 The Bronze Horses of St. Mark's.
7 Interior of St. Mark's Cathedral.
8 The Rood Screen, St. Mark's.
9 St. Mark's Cathedral, from the Ducal Palace.
10 Bronze Wells, Ducal Palace.
11 Ducal Palace, the Inner Facade.
12 Tame Pigeons of St. Mark's.
13 The Giant Stairs, Ducal Palace.
14 Ducal Palace, the Canal Front.
15 The Bridge of Sighs.
16 The Vine Angle of the Ducal Palace.
17 The Granite Columns on the Piazzetta.
18 The Winged Lion of St. Mark's.
19 The Riva degli Schiavoni.
20 A Venetian Street, and Leaning Campanile.

21 St. Zaccaria Church.
22 The Arsenal, entrance.
23 The Public Gardens, Venice.
24 St. Giorgio from the Piazzetta.
25 Venice from St. Giorgio Campanile.
26 Eastern Venice from St. Giorgio.
27 The Cavalli Palace, Grand Canal
28 The Grand Canal, from the Iron Bridge.
29 Church of S. Maria della Salute
30 Quay and Canal from della Salute Church.
31 Group of Women at a Well.
32 The Foscari Palaces, Grand Canal.
33 The Ponte Rialto.
34 Grand Canal, from Ponte Rialto.
35 Grand Canal, South of the Rialto.
36 The Market, from Ponte Rialto.
37 Grand Canal, from the Turkish Palace.
38 Church of Gli Scalzi, interior.
39 A Paved Street, "Calle del Sturion."
40 Via Alla Posta, Street at the Post-Office.

All the Negatives for the above Lecture sets were taken in the Summer of 1884

QUEEN & CO., PHILADELPHIA.

Series IV.—American History.

PLAIN PHOTOGRAPHS, 50 CENTS; COLORED, 3 INCHES DIAMETER, $1.50 :

No.
1 Landing of Columbus, 1492.
2 Smith rescued by Pocahontas, 1607.
3 Marriage of Pocahontas, 1613.
4 Embarkation of Pilgrim Fathers, 1620.
5 Landing of the Pilgrims, 1620.
6 Landing of Roger Williams, 1630.
7 Roger Williams sheltered by Indians, 1636.
8 Penn's Treaty with the Indians, 1682.
9 Retreat of Braddock, 1775.
10 Washington raising the British Flag at Fort Duquesne, 1758.
11 Boston Massacre, 1775.
12 The Boston Tea Party, 1775.
13 Putnam Leaving the Plough, 1775.
14 Struggle on Concord Bridge, 1775.
15 Battle of Lexington, 1775.
16 Battle of Bunker Hill, 1775.
17 Washington taking Command of the Army, 1775.
18 Capture of Fort Ticonderoga, 1775.
19 Putnam's Escape, 1775.
20 Drafting the Declaration of Independence, 1776.
21 Declaration of Independence, 1776.
22 Lord Sterling at Battle of Long Island, 1776.
23 Defense of Fort Moultrie, 1776.
24 Washington Crossing the Delaware, 1776.
25 Battle of Harlem, 1777.
26 Surrender of Burgoyne, 1777.
27 Washington at Valley Forge, 1777.
28 Moll Pitcher at Monmouth, 1778.
29 Indian Massacre at Wyoming, 1778.
30 Battle of Stony Point, 1779.
31 Gen. Marion and British Officer,1780.

No.
32 Capture of Andre, 1780.
33 Surrender of Cornwallis, 1781.
34 Commodore Perry at Lake Erie, 1813.
35 Battle of New Orleans, 1815.
36 General Scott entering Mexico, 1847.
37 Bombardment of Fort Sumter, April 12, 1861.
38 Battle of Bull Run, July 16–19, 1861.
39 Battle of Wilson's Creek, Aug. 9, 1861.
40 Battle of Roanoke Island, Feb.8, 1862.
41 Capture of Fort Donelson.
42 Battle of Pittsburg Landing.
43 Capture of New Orleans, Ap. 25, 1862.
44 Battle of Fair Oaks, May 31, 1862.
45 Battle of Antietam, Sept. 17, 1862.
46 Attack on Fredericksburg, Dec. 13, 1862.
47 Siege of Vicksburg, July, 1863.
48 Battle of Gettysburg, July 1–3, 1863.
49 Battle of Chickamauga, Sept. 19-20, 1863.   [1863.
50 Battle of Lookout Mountain, Nov. 24,
51 Battle of the Wilderness, May 5-6, 1864.
52 Attack on Fort Wagner.
53 Capture of Petersburg, Ap. 2, 1865.
54 Naval Combat between Monitor and Merrimac, March 9, 1862.
55 Naval Combat between Kearsarge and Alabama.
56 Surrender of Gen. Lee, Ap. 9, 1865.
57 First Reading of Emancipation Proclamation.
58 Sherman's March through Georgia, December, 1864.
59 Assassination of Abraham Lincoln, April 14, 1865.
60 Capture of Jeff. Davis, May 10, 1865.

American Civil War.

Battle of Malvern Hill.
Battle of Newburn.
Bombardment of Port Royal.
Battle of Williamsburg.
Attack on Kelly's Ford, Va.
Attack on Port Hudson.
Battle of Mill Creek.
Battle of Shiloh—Gen. Grant's Charge.
Attack of Gunboat at Memphis.
Battle of Rich Mountain.
Battle of Murfreesboro'.
Massachusetts Troops passing through Baltimore.

Battle of Chantilly—Gen. Kearney's Charge.
Battle of Ball's Bluff.
Bombardment of Island No. 10.
Bombardment of Mobile Bay.
Assassination of Ellsworth.
Ellsworth Revenged.
Capture of Atlanta.
Sherman's Army entering Savannah.
Gettysburg.
Lookout Mountain.
Fredericksburg.

Illustrations of the Manners and Customs of the North American Indians.

PLAIN PHOTOGRAPHS, 50 CENTS; COLORED, 3 INCHES DIAMETER, $1.50

No.
1. Red Jacket.
2. The Death Whoop.
3. Combat between the Ojibewas and the Sacs and Foxes.
4. Emigrants attacked by the Comanches.
5. The Dog Dance of the Dacotahs.
6. Indian Council.
7. Hunting Buffalo in Winter.
8. Gathering Wild Rice.

No.
9. Shooting Fish.
10. Indian Women procuring Fuel.
11. Indian Doctor preparing a Pot of Medicine.
12. Indian Seer attempting to destroy a female with Enchanted Sunbeams.
13. Indian Burial.
14. Nocturnal Grave Light, Indian Girl Watching by the Grave.

Set of Twenty Illustrations of Ancient Greece.

Beautifully Colored Photographs of Fine Engravings.

1. Plan of Athens.
2. Ancient Athens, Restored.
3. Ruins of Athens.
4. The Piræus.
5. Mars' Hill.
6. The Philosopher's Garden.
7. Ruins of the Parthenon.
8. The Parthenon, Restored.
9. Temple of Jupiter Olympus.
10 Oracle at Delphia.

11. Sacrifice to Neptune.
12. " to Mars.
13. Statue of Pallas Athenæ.
14. Olympic Games.
15. Grecian Warriors.
16. " Chariot.
17. " Dwelling, Interior.
18. " Ceremony before Marriage
19. The Areopagus.
20. The Assembly of the Gods.

Egyptian Antiquities.

1. Egyptian Wine Press.
2. " Capitals.
3. " Royal Boat.
4. " Ancient Armor.

5. Egyptian Harper, from Belzoni's Tomb.
6. Egyptian Chair, from Tomb of Rameses.

A variety of other subjects.

Assyria.

1. Mount Ararat.
2. Birs Nimroud.
3. Source of Tigris.
4. Urfah.
5. Hamadan and Ruins of Castle Darius.
6. Ruins of Persepolis.
7. Interior of a Caravansera.

8. Great Mosque of Urfah.
9. Ruins of Babylon.
10. " Sus.
11. " Nineveh.
12. " Palmyra.
13. Palace of the Cæsars
14. City of Aleppo.
15. " Damascus.

Set of Twenty Illustrations of Ancient Rome.
Beautifully Colored Photographs of Fine Engravings.

1. Map of Rome.
2. Ruins   "
3. Trajan's Arch.
4. Roman Cavalry.
5. War Elephant.
6. " Engines.
7. Victorious General Thanking his Army.
8. Prisoners passing under the Yoke.
9. Roman Triumph.
10. Captives in the Forum.
11. Gladiators at the Theatre.
12. " Funerals.
13. Sea Fight.
14. Roman Feast.
15. The Colosseum.
16. Section of the Colosseum.
17. Wild Beasts and Victims in the Colosseum.
18. Sacrifice in Rome.
19. Temple of the Sun in Rome.
20. Funeral of an Emperor.

Solomon's Temple and the Tabernacle in the Wilderness.
Illustrations from Paine's Book.
SOLOMON'S TEMPLE.

1. Side and Interior View of a Gate of the Temple.
2. Gates of Solomon's Temple.
3. North View of Solomon's Temple.
4. Ground-plan " "
5. Solomon's Temple, East View.
6. " " West "
7. " " Front "
8. Interior of Solomon's Temple.
9. Ground-plan " "
10. The Tabernacle.
11. Interior of Tabernacle.
12. Tabernacle and Court.
13. Planks of the Tabernacle.
14. High Priest and Candlestick.
15. House of the King, Side View.
16. " " " Interior.
17. Pavements of Solomon's Temple
18. Herod's Temple.
19. " " Gate and Substructures.

THE TABERNACLE IN THE WILDERNESS.

1. The Tabernacle and Camp.
2. Holy Place and Most Holy.
3. High Priest in Linen Robes.
4. " in "Garments of Beauty."
5. Brazen Altar and Covering.
6. Candlestick and Covering.
7. Ark " "
8. Altar of Incense and Covering.
9. Brazen Laver.
10. Table of Showbread.

53

Series V.—Bible Illustrations.

Beautifully Colored. From Designs by Neslie.

PLAIN PHOTOGRAPHS, 50 CENTS; COLORED, 3 INCHES DIAMETER. $1 50.

NO.
1. The earth without form and void. Gen. i, 3, 4.
2. Waters gathered in one place—the dry land appears. Gen. i, 9.
3. The earth yields grass and fruit trees. Gen. i, 11.
4. God places two great lights, and stars also. Gen. i, 16, 17.
5. God creates the fowls and fish. Gen. i, 20.
6. God creates cattle, creeping things, and beasts. Gen. i, 24.
7. God creates man and gives dominion. Gen. ii, 7.
8. God forms woman from the rib of man. Gen. ii, 22.
9. The woman, being tempted, eats. Gen. iii, 6.
10. The woman accuses the Serpent of beguiling her. Gen. iii, 13.
11. God drives Adam and Eve from the Garden. Gen. iii, 24.
12. Adam, Eve, Cain, and Abel, the first human family. Gen. iii, 19.
13. Cain's offering rejected. Gen. iv, 3-5.
14. Cain kills his brother Abel. Gen. iv, 8.
15. The curse of Cain. Gen. iv, 11, 12.
16. Cain builds the first city. Gen. iv, 17.
17. The Three Tribes descended from Cain. Gen. iv, 20-23.
18. The wickedness of mankind before the Flood. Gen. vi, 5, 6.
19. God commands Noah to build the Ark. Gen. vi, 14-22.
20. The Flood destroying man and beast. Gen. vii, 23.
21. The interior of the Ark. Noah and his family surrounded by animals.
22. The Dove coming to Noah with the olive branch. Gen. viii, 11.
23. Noah's Sacrifice. Gen. viii, 20.
24. The scattering of the Tribes from Babylon upon the face of the earth. Gen. xi, 7, 8.
25. Destruction of the Cities of the Plain. Gen xix, 24-26.
26. Joseph thrown into the well. Gen. xxxvii, 24.
27. Joseph sold by his Brethren to the Midianites. Gen. xxxvii, 28.
28. Joseph's bloody coat shown to Jacob. Gen. xxxvii 33, 34.
29. Joseph interprets the Dreams of the Butler and Baker. Gen. xl, 8.
30. Joseph interprets Pharaoh's Dream. Gen. xli, 16.
31. Simeon detained by Joseph. Gen. xlii, 24.
32. Joseph makes himself known to his Brethren. Gen. xlv, 3.
33. Joseph meets his Father in Goshen. Gen. xlvi, 29.
34. Jacob blesses his Twelve Sons. Gen. xlix, 2.
35. The Angel of the Lord appears to Moses. Exod. iii, 2.
36. Pharaoh and his hosts drowned in the Red Sea. Exod. xiv, 28.
37. Moses strikes the Rock at Rephidim. Exod. xiv, 6.
38. Moses receives the Tablets at Mount Sinai. Exod. xxxi, 18.
39. Falling down of the Walls of Jericho. Josh. vi, 20.
40. Gideon defeats the Midianites with Lamps and Trumpets. Judges vii, 20.
41. Samson and the Lion. Judges xiv, 5.
42. Samson betrayed by Delilah. Judges xvi, 29.
43. Samson grinds corn in the Prison-house. Judges xvi, 21.
44. Samson destroying the Temple. Judges xvi, 29.
45. David slaying Goliath. 1 Sam. xvii, 49.
46. The raising of Samuel by the Witch of Endor. 1 Sam. xxviii, 11.
47. Absalom entangled in the Oak. 2 Sam, xviii, 9.
48. Elijah ascending to Heaven in the presence of Elisha. 2 Kings ii, 11.
49. David bringing the Ark from Kirjath-Jearim. 2 Sam. vi, 5.
50. Children in Fiery Furnace. Daniel iii, 23.
51. Daniel in the Lions' Den. Daniel vi, 16.
52. Jonah cast into the Sea. Jonah i, 16.
53. Jacob's Dream. Gen. xxviii, 11.

No.
54. Moses delivering the Law to the people. Exod. xix, 25.
55. The Golden Calf. Exod. xxxii, 19.
56. Jeremiah weeping over Jerusalem. Jer. xxxix, 8.
57. The Death of Abel.
58. Abraham offering Isaac.
59. Hagar and Ishmael in Desert.
60. Jacob's Dream.

PLAIN PHOTOGRAPHS, 50 CENTS; COLORED, 3 INCHES DIAMETER, $1.50.

ADDITIONAL OLD TESTAMENT.

61. Adam and Eve in Paradise. *Gosse.*
62. The Sacrifice of Noah. *Maclise.*
63. Rebecca at the Well. *Schopin.*
64. Eleazar in the House of Bethuel. *Schopin.*
65. Arrival of Rebecca. *Schopin.*
66. Jacob waters the Flocks of Rachel. *Glaize.*
67. Joseph sold by his Brothers. *Schopin.*
68. Joseph's bloody coat brought to Jacob. *Schopin.*
69. Joseph meets his Father in Goshen. *Schopin.*
70. Moses saved by Pharaoh's Daughter. *Schopin.*
71. Moses assisting the Daughters of Jethro. *Schopin.*
72. Pharaoh's host drowned in Red Sea. *Schopin.*
73. Jephthah's Daughter meeting her Father. *Glaize.*
74. Samson betrayed by Delilah. *Schopin.*
75. David returns Conqueror of Goliath. *Schopin.*
76. David in camp of Saul. *Schopin.*
77. Saul and the Witch of Endor. *Alston.*
78. The Judgment of King Solomon. *Schopin.*
79. Solomon's Reception of Queen of Sheba. *Schopin.*
80. Espousal of Esther by Ahasuerus. *Schopin.*
81. Esther implores Ahasuerus. *Schopin.*
82. Jeremiah weeping over Jerusalem. *Bendeman.*
83. The Feast of Belshazzar. *Schopin.*
84. Daniel in the Lions' Den. *Zeigler.*

LIFE OF OUR SAVIOUR.

85. The Annunciation of the Virgin. *Jalabert.*
86. The Angel appearing to the Shepherds. *White.*
87. Holy Night. *Correggio.*
88. The Magi guided by the Star. *Portaels.*
89. Adoration of Magi. *Veronese.*
90. The Presentation in the Temple. *Dowling.*
91. The Flight into Egypt. *Portaels.*
92. The Shadow of the Cross. *Morris.*
93. The Return to Nazareth. *Dobson.*
94. Jesus disputing with the Doctors. *Hunt.*
95. Miraculous Draught of Fishes.
96. The Baptism of Christ. *Wood.*
97. Christ tempted by the Devil. *Scheffer.*
98. Christ and the Samaritan Woman. *Herbert.*
99. Christ preaching on Sea of Galilee. *Hoffman.*
100. The Sermon on the Mount. *Debufe.*
'01. Christ healing the Sick. *West.*
102. Christ raising the daughter of Jarius. *Richter.*
103. Christ walking on the Waters. *Jalabert.*
104. The Miracle of the Loaves. *Dubufe.*
105. The Transfiguration. *Raphael.*
106. Parable of Prodigal Son, Carousal. *Dubufe.*
107. Parable of Prodigal Son, Swineherd. *Dubufe.*
108. Parable of Prodigal Son, Return. *Dubufe.*
109. Christ blessing the Little Children. *Bida.*
110. Mary Magdalen washing the feet of Jesus. *Barrias.*
111. Christ and the Rich Young Man. *Lejeune.*
112. The Parable of the Lilies. *Lejeune.*
113. Christ, the Outcast of the People. *Herbert.*
114. Christ's Entry into Jerusalem. *Dubufe.*
115. The poor Widow's Two Mites. *Barrias.*
116. Christ's Charge to Peter. *Raphael.*

No.
117. Christ weeping over Jerusalem. *Eastlake.*
118. The Last Supper. *De Vinci.*
119. The Agony in the Garden. *Scheffer.*
120. Christ before Pilate. *Munkacsy.*
121. Christ bearing the Cross. *Veronese.*
122. Christ arriving at Mount Calvary *Steuben.*
123. The Crucifixion. *Munkacsy.*
124. Golgotha, "It is finished." *Gerome.*

No.
125. The Descent from the Cross. *Rubens.*
126. The body of Christ laid in the Tomb. *Jalabert.*
127. The Marys at the Tomb. *Schleh.*
128. The Resurrection of Christ. *Bamberg.*
129. First Easter Dawn. *Thomson.*
130. Easter Morning. *Ploekhorst.*
131. The Journey to Emmaus. *Roberts.*
132. The Ascension of Christ. *Gleyre.*

Stations of the Cross.

Station 1.—Jesus condemned to Death.— S. Luke xviii, 24.
Station 2.—Jesus is laden with the Cross. —S. John xix, 17.
Station 3.—Jesus falls the first time under the weight of the Cross.—Isaiah liii, 7
Station 4.—Jesus meets His most Holy Mother.—S. John ii, 1.
Station 5.—Jesus is helped by the Cyrean to carry His Cross.—S. Luke xxiii, 26.
Station 6.—Veronica wipes the face of Jesus.—Psalm xxvii, 8.
Station 7.—Jesus falls beneath His Cross the second time.—Psalm xxxviii, 7.

Station 8.—Jesus consoles the Women of Jerusalem.—S. Luke xxiii, 28.
Station 9.—Jesus falls beneath His Cross the third time.—Psalm xxii, 15.
Station 10.—Jesus is stripped of His garments, and is given gall to drink. —Psalm xxii, 15.
Station 11.—Jesus is nailed to the Cross. —S. Luke xxiii, 33.
Station 12.—Jesus is raised on the Cross and dies upon it.—S. John xix, 30.
Station 13.—Jesus taken down from the Cross.—S. Mark xv, 46.
Station 14.—Jesus laid in the Holy Sepulchre.—S. John xix, 42.

Life of Moses.

1. Striking the Rock.
2. Jethro bringing his Wife and Children.
3. Coming down from the Mount.
4. Destroying the Golden Calf.
5. The Rebellion of Korah.
6. Lifting up the Serpent.
7. Spoiling of the Midianites.
8. Viewing the Promised Land from Pisgah.
9. Finding of Moses.
10. Slaying the Egyptian.
11. The Burning Bush.
12. Returning to Egypt.
13. Before Pharaoh.
14. Plague of Locusts.
15. Slaying the First-born.
16. The Departure from Egypt.
17. The Red Sea
18. Miriam.

PILGRIM'S PROGRESS.

From "The Art Journal" Illustrations.

1 The Genius of Art delineating Bunyan's Dream.
2 Christian meditates his Departure.
3 Christian is met by Evangelist.
4 Pliable consents to bear Christian company.
5 Christian and Pliable fall into the Slough of Despond.
6 Christian's Danger beneath Mount Sinai
7 Christain is released from his Burthen of Sin.
8 Christian endeavors to awake Sloth, Ignorance and Presumption.
9 Christian is met by Fear and Mistrust.
10 Christian combats with Apollyon.
11 Christian vanquishes Apollyon.
12 Christian prepares to enter the Valley of the Shadow of Death.
13 The Meeting of Christian and Faithful.
14 Christian and Faithful mocked by the Scorners of Vanity Fair
15 The Destruction of By-ends and his Companions.
16 Christian and Hopeful are seized by the Giant Despair.
17 Christian and Hopeful Escape from the Giant Despair.
18 Christian and Hopeful are shown the Entrance to the Bottomless Pit.
19 Christian and Hopeful behold the Fate of the Apostate.
20 Christian and Hopeful arrive at the Waters of Death.
21 Christian and Hopeful pass the Waters of Death.
22 Christian and Hopeful ascend into the Celestial City.

CURFEW MUST NOT RING TO-NIGHT.

Reading by Rosa Hartwick Thorpe. $10.00.

No.
1. Introductory Design, with Title.
2. Country Landscape, with Sunset Effect.
3. Bessie walking with the Sexton.
4. The Sexton refuses Bessie's request.
5. Bessie climbing the Belfry stairs.
6. Bessie reaches the Bell.

No.
7. Bessie clinging to the Bell and swinging to and fro (Lever Slide).
8. The Sexton Ringing the Bell.
9. Bessie pleading to Cromwell to spare her Lover.
10. Release of Basil Underwood. The Lovers' meeting.

Little Jim the Collier Boy.

BY E. FARMER. WITH LECTURE, $3.00.

No.
1. The Cottage was a thatched one.
2. With hands uplifted, see, she kneels.
No.
3. With gentle, trembling haste, she held.
4. The Cottage door was opened.
No.
5. He knew that all was over.
6. His quivering lips gave token.

57

Uncle Tom's Cabin.
AND READING.

No.
1. George Harris taking leave of his Wife.
2. An Evening in Uncle Tom's Cabin.
3. Escape of Eliza and Child on the ice.
4. Uncle Tom sold and leaving his family.
5. Eva St. Clair makes a friend of Uncle Tom.
6. Uncle Tom saves Eva from drowning.

No.
7. George Harris resisting the Slave Hunters.
8. Eva and Topsy.
9. Eva reading to Uncle Tom.
10. Eva's dying farewell.
11. Legree's Cruelty to Uncle Tom.
12. Death of Uncle Tom.

Rip Van Winkle.
With Reading.

BY WASHINGTON IRVING. SEVEN ILLUSTRATIONS OF THE ROMANCE, AND FIVE VIEWS FROM NATURE OF THE CATSKILL MOUNTAINS, THE HAUNTS OF RIP VAN WINKLE.

No.
1 Dame Van Winkle expressing her mind
2 Rip Van Winkle's perfect contentment
3 Derrick Van Bummil reading the news

No.
4 Rip Van Winkle waiting upon the strange company
5 Rip Van Winkle after his long sleep
6 Rip Van Winkle relating his story
7 The Haunts of Rip Van Winkle

No.
8 Sunset Rock, Catskill
9 The Catskill Water Fall
10 The Cascades of the Catskill
11 Bastion Falls, Catskills
12 Cascade Profile Rock

Life of Washington.
Finely Colored.

1 The Cherry Tree Incident.
2 Young Washington as a Peacemaker.
3 Courtship of Washington.
4. Washington Crossing the Delaware.

5. The Prayer at Valley Forge.
6. The Inaugural Address of Washington
7. Lafayette at Mount Vernon.
8. Last Moments of Washington.

Series X.—INTEMPERANCE.
Ten Nights in a Bar-Room.

1. The arrival at the "Sickle and Sheaf."
2. Joe Morgan's little Mary begs him to come home.
3. Slade throws a glass at Joe Morgan and hits Mary.
4. Joe Morgan suffering the horrors of Delirium Tremens.
5. Death of Joe Morgan's little Mary.
6. Frank Slade and Tom Wilkins riding off on a spree.
7. Willie Hammond is induced by Harvey Green to gamble.
8. Harvey Green stabs Willie Hammond to death.
9. Quarrel between Slade and his son Frank.
10. Frank Slade kills his father with a bottle.
11. Meeting of the Citizens in the Bar-Room.
12. The Departure from the "Sickle and Sheaf."

The Stomach of the Drunkard in its different Stages of Disease.

PER SLIDE, $1.50.

No.
1. Appearance of the Stomach of a Temperance Man.
2. Appearance of the Stomach of the Moderate Drinker.
3. Appearance of the Stomach of a Drunkard.
4. Appearance of the Stomach after a Debauch.

No.
5. Appearance of the Stomach of a Hard Drinker.
6. Appearance of the Stomach of a Habitual Drunkard.
7. Appearance of the Stomach of a Drunkard on the verge of the grave.
8. Appearance of the Stomach during Delirium Tremens.

The Drunkard's Career and End.

Beautifully Colored Photographs of Fine Engravings.

PER SLIDE, $1.50.

1. Domestic Happiness. The greatest of earthly blessings.
2. The Temptation. "Lead me not into temptation."
3 Introduction of Sorrow. A loving heart made sad.
4. The Rum-hole. A substitute for home.
5. Rum instead of Reason.
6. Degraded Humanity.

7. The Cold Shoulder by Old Friends.
8. Rumseller's Gratitude. Rejection instead of Injection.
9. Poverty and Want.
10. Robbery and Murder. The result of drunkenness.
11. Mania-a-potu. The horror of horrors.
12. The Death that precedes Eternal Death.

The Bottle.

Eight Beautifully Colored Photographs of Fine Engravings.

From the Originals, by G. Cruikshank.

PER SLIDE, $1.50.

1. The bottle is brought out for the first time. The husband induces his wife "just to take a drop."
2. He is discharged from his employment for drunkenness. "They pawn their clothes to supply the bottle."
3. An execution sweeps off the greater part of their furniture. "They comfort themselves with the bottle."
4. Unable to obtain employment, they are driven by poverty into the streets to beg, and by this means still supply the bottle.

5. Cold, misery, and want destroy their youngest child. "They console themselves with the bottle."
6. Fearful quarrels and brutal violence are the natural consequences of the frequent use of the bottle.
7. The husband, in a state of furious drunkenness, kills his wife with the instrument of all their misery.
8. The bottle has done its work. It has destroyed the infant and the mother; it has brought the son and daughter to vice and to the streets, and has left the father a hopeless maniac.

New Temperance Set—Matt Stubbs' Dream. (*Colo'd only.*)

$12.00.

1 Title.
2 There came a woman in with a baby.
3 He was leaning over the back of a chair
4 Opened the door letting out a glow of light
5 Kate's gentle voice went through the chapter
6 Looking down on a wee bit of humanity
7 "Rum!" said Matt, opening his eyes
8 There was every class, from the king to squalid rags
9 Drape! from head to foot in black
10 "Oh! please don't," cried the boy
11 Grasping the hand-rail to steady himself
12 "Uncle tell me a story," said the little tyrant
13 Said with all her heart "God bless you, Matt"

Sam Bowen's Dream. (*Colo'd only.*)

$10.00.

A Temperance Story, *from original designs*.

1 Introduction
2 A bleak afternoon in Feb'ry
3 As sulky as a bear
4 And then he had a row
5 The poor brute bolted and upset the trap
6 It was one of the first days of harvest
7 She helped me up-stairs
8 The room seemed queer and foggy
9 "Gone," says my wife
9a Effect
10 In the garden
10a Angel effect
11 Welcome home
12 The good old Bible

With Reading.

A Christmas Hymn.

With Reading.

1 "Had Rome been growing up to might, and now was queen of land and sea"
2 "The Senator of haughty Rome impatient urged his chariot's flight"
3 "Within that province far away went plodding home a weary boor"
4 "How calm a moment may precede one that shall thrill the world forever"
5 "A thousand bells ring out, and throw their joyful peals abroad"
6 "For in that stable lay, new born, the peaceful Prince of earth and heaven"

The Rock of Ages.

1 The Angry Sea
2 The Rock of Ages
3 Flashes of Lightning
4 Rainbow
5 The Helping Hand
6 Simply to thy Cross I cling
7 Angels Beckoning
8 Angel Crowns Faith
9 Ascension to Heaven
10 Heaven

Nos. 3, 4, 7, and 8 can only be used with two Lanterns for Dissolving

New Tale of a Tub.

Finely Executed Pictures.—A Comic Poem Illustrated.

1 Opening the Question. The Bengal Tigers
2 Bengal Cave
3 The Artful Dodge
4 Look before you leap
5 Under Cover.
6 Increasing the Interest of the Tail

The Gambler's Career.

Beautifully Colored Photographs of Fine Engravings.

1 The first seed of the passion planted in the young mind
2 The development of the passion with higher stakes
3 Finding himself always the loser, he resorts to false play
4 He is detected and roughly handled by his friends
5 Having finally lost his all, he leaves the gambling-house in despair and madness
6 He ends his life in a mad house, still occupied with his ruling passion

61

SECRET SOCIETIES.

Series XII.—Masonic.

Beautifully Colored.

FIRST DEGREE.

1 Holy Bible, Square, Compass, and Warrant
2 Ancient Lodge in Valley
3 Form of Lodge
4 Supports of Lodge
5 Jacob's Ladder
6 Furniture of Lodge
7 Ornaments of Lodge
8 Lights of Lodge
9 Jewels of Lodge
10 Tabernacle in Wilderness.
11 St. John the Baptist, and St. John the Evangelist
12 Masonic Tenets
13 Points of Entrance
14 Chalk, Charcoal, and Clay.

FELLOWCRAFT'S DEGREE.

15 Pillars of the Porch
16 Five Orders of Architecture
17 The Five Senses
18 Seven Liberal Arts
19 Scene at the Waterfall
20 Corn, Wine, and Oil
21 Allusion to the Letter G

MASTER MASON'S DEGREE.

22 Building of King Solomon's Temple
23 Marble Monument
24 Ancient Three Grand Masters
25 Entered Apprentice's Lodge
26 Fellowcraft's Lodge
27 Master Mason's Lodge
28 Three Steps
29 Pot of Incense
30 Bee-Hive
31 Book of Constitution Guarded by Tyler's Sword
32 Sword Pointing to Naked Heart, and All-Seeing Eye
33 Anchor and Ark
34 Forty-seventh Problem
35 The Hour-Glass
36 The Scythe
37 Emblems of Mortality

ROYAL ARCH CHAPTER.

25. The Burning Bush.

Commandery of Knights Templar.

26 Angel at Sepulchre.   27. Three Marys at Tomb.   28. Ascension of Christ

29. Valley of Dry Bones.
30. The Crucifixion.
31. Body of Christ lain in Tomb.
32. Resurrection of Christ.
33. The Cross.
34. The Pilgrim.
35. The Knight.
36. The Penitent.

37. Christ on the Cross.
38. Death on the Pale Horse.
39. Human Skull.
40. John at Patmos.
41. Faith at the Cross.
42. Cross and Crown of Glory, with motto—"Crown of Life."

(No. 28, Perpendicular Movement, $6.50.)

Series XIII.—Odd Fellows.

A new and Superior Series, from New Designs, for the New Work of the INDEPENDENT ORDER OF ODD FELLOWS. Per Slide, $1.50.

Initiatory Degree.—1. All-seeing Eye. 2. Three Links. 3. Skull and Cross Bones. 4. The Scythe.

First Degree.—5. Bow and Arrow. 6. The Quiver. 7. The Bundle of Sticks.

Second Degree.—8. The Axe. 9. Heart and Hand. 10. The Globe. 11. The Ark. 12. The Serpent.

Third Degree.—13. Scales and Sword. 14. The Bible. 15. The Hour Glass. 16. The Coffin.

ENCAMPMENT EMBLEMS. — 17. The Three Pillars. 18. The Tent. 19. The Pilgrim's Scrip, Sandals, and Staff. 20. The Altar of Sacrifice. 21. The Tables of Stone, Crescent and Cross. 22. The Altar of Incense.

P. O. S. of America.

1 Gallileo Expounding his Theories.
2 Gallileo before the Inquisition.
3 Columbus Discovering America.
4 The Mayflower.
5 Landing of the Pilgrims.
6 Battle of Lexington.
7 Battle of Bunker Hill.
8 Portrait of Washington
9 Washington Crossing the Delaware.
10 Washington at Prayer at Valley Forge.
11 Battle of Bennington.
12 Battle of Saratoga.
13 Battle of Monmouth.
14 Battle of Stony Point.
15 Battle of Cowpens.

16 Battle of Eutaw Springs.
17 Surrender of Cornwallis at Yorktown.
18 Battle of New Orleans.
19 Hall of the Montezumas.
20 Firing on Fort Sumter.
21 Rally of Troops at Washington.
22 Battle of Gettysburg.
23 Battles of the Civil War (as many as desired).
24 Surrender of Lee.
25 Scene of Peace.
26 Public School House.
27 Goddess of Liberty.
28 Stars and Stripes.

Knights of Pythias.

First Rank.

1 The Friends, Damon and Pythias.
2 Damon Condemned to Die.
3 Pythias' Appeal to Dionysius.
4 The Flight of Damon to his Family.
5 Pythias in Dungeon, Calantha's Appeal.
6 Damon's Farewell to his Family.
7 Pythias at the Headsman's Block.
8 Pythias Saved by Damon's Arrival.
9 The Heroes Honored by the King.
10 The Beautiful Unknown Shore.

Third Rank.

1 Arts of Ancient Egypt.
2 A Flower Bespangled Plain.
3 The Mountain Side.
4 A Sunless Sea.
5 Where Hideous Creatures Climb.
6 The Hero.

Sixth Senator.

1 The Deadly Conflict.
2 The Fallen Soldier.
3 The Wounded Soldier Assisted.

NOTE.—This set can be *furnished in colored slides only.*

Series XV.—Comic Subjects.

PLAIN, 50 CENTS EACH.

No.
1. A Lovely Calm, No. 1.
2. A Black Squall, No. 2.
3. "Come into the Garden, Maud."
4. "We Met by Chance."
8. "'Twere vain to tell Thee all I Feel."
9. "Darling, I am Growing Old."
10. "'Twas a Calm, Still Night."
12. "Angel Voices Sweetly Calling."
13. Laying back Stiff for a Brush, No. 1.
14. Hung up, with the Starch Out, No. 2.
15. The Parson's Colt Trots if it is Sunday.
16. Between Two Fires.
17. A Bare Chance.
20. A Game Dog.
21. Boss of the Road.
22. Bull-dozed.
27. Man as He Expects to be.
28. "Triumph of Woman's Rights."
29. Five Degrees of Intemperance.
33. The Summit of Happiness. No. 1.
34. The Depth of Despair. No. 2.
35. Walked Home on his Ear.
36. Life in Death.
37. Trouble in de Church. No. 1.
38. "   "   "   No. 2.
39. "Go way down dar." No. 1.
40. "Dar, I knew Mischief was Brewin'." No. 2.
41. The Attack on the Watermelon.
42. Profit and Loss.
44. "Coming Thro' the Rye."
45. We've had a Healthy Time.
47. "In Happy Moments." No. 1.
48. Star of the Evening, Gently Guide Me. No. 2.
51. Golly, no wonder Missus don't get up till 10 o'clock.
52. Venus Rising from the Sea.
53. She Stoops to Conquer.
54. "I want to be an Angel."
55. "Put my Little Shoes Away."
56. Ignorance is Bliss.
57. Something has got to come.
58. "Shimply (hic) Waitin' for a Fren'."
59. "They All Do It."
60. O Where, O Where, has my Leedle Dog Gone?
61. What is Home Without a Mother-in-Law?
64. Dot Leedle German Band.
65. What are the Wild Waves Saying, Sister?" No. 1.

No.
66. "Scoot, Brother, Scoot." No. 2.
67. "Listen to the Mocking Bird."
68. "That Husband of Mine."
69. "Stolen Pleasures are Sweet." No. 1.
70. "No Pleasure Without Pain." " 2.
73. "Babies on our Block."
75. Injured Innocence.
76. Moving Day.
78. Fatherless.
80. Three Systems of Medicine.
81. "Excuse Haste and a Bad Pen."
82. A Division of Labor.
83. Maternal Solicitude; Monkeys.
84. Going! Going! Gone!
85. Pleasure Before Business
87. The Finding of Moses, by Titian.
88. "   " by Mark Twain.
89. Two Souls with but a Single Thought.
90. Don't You Forget It.
91. A Mule Train on an Up Grade.
92. "   " on a Down Grade.
95. "Sure of a Bite."
96. "Bustin' a Picnic."
109. "Mary Had a Little Lamb."
110. "The Girl I Left Behind Me."
111. "Too Late for the Train."
112. First Lesson in the Battle of Life.
113. The Chinese Question—The Rivals.
114. "   " The Controversy Settled.
115. A Ghost Adventure—A Moonlight Reverie.
116. A Ghost Adventure—A Ghostly Problem.
117. A Ghost Adventure—The Problem Solved.
118. The Pre-Historic Fop, According to Darwin.
119. The Modern Fop, According to the 15th Amendment.
120. Attack of the Monster—The Wicked Flea.
121. Attack of the Monster—Boarding-house Bedbug.
122. How Jones Became a Mason—Starting for the Lodge.
123. How Jones Became a Mason—Oath of Secrecy.
124. How Jones Became a Mason—Riding the Goat.
125. How Jones Became a Mason—Jones is a Mason
126. The Schoolboy's First Cigar.
127. "   " " Its Result.

No.
128. The Onconvanience of Single Life.
129. The Raal Convanience of Married Life.
130. The Masher.
131. " Crushed.
132. I Wonder if 'tis Loaded ?
133. It Was Loaded!
134. Joy!
135. Horror!
136. The Little Peach—**Expectation.**
137. " " **Realization.**
138. " " **Termination.**
139. Peace.
140. War.
141. Why Did You Sup on Pork?
142. A Coolness Between Friends.
143. Mr. Murphy is Rising in the World.

No.
144. A Pleasure Party.
145. Two Heads Are Better Than One
146. War Dance—Opening of the Ball
147. The Three Graces.
148. Richard is Himself Again.
149. The Three (Scape) Graces.
152. Victory Doubtful.
157. Something Did Come!
158. The Pet of the **Ladies—The Exquisite.**
159. The Pet of the Fancy—The Prize Fighter.
166. "Every Dog Has His Day."
167. "Thou Art So Near and Yet So Far."
168. "Grab the Ball, Johnny, I'll Wait."
169. **Great Expectations.**

Series XVI.—Choice Statuary.

Photographs from the Originals, Backgrounds Blacked. Mounted in Wood Frames.

By JOHN ROGERS, the Famous American Sculptor.

PER SLIDE, $1.00, FRAMED; 75 CENTS UNFRAMED.

1. Fairy's Whisper.
2. Fugitive's Story.
3. Council of War.
4. Challenging the Union Vote.
5. Taking the Oath.
6. The Favored Scholar.
7. The Foundling.
8. Coming to the Parson.
9. Courtship in Sleepy Hollow.
10. One More Shot.
11. Wounded Scout.
12. Union Refugees.
13. Country Post-office.
14. Home Guard.
15. School Examination.
16. Charity Patient.
17. Uncle Ned's School.
18. Returned Volunteer.
19. Playing Doctor.
20. Parting Promise.
21. Rip Van Winkle at Home.
22. Rip Van Winkle on the Mountain.
23. Rip Van Winkle Returned.
24. We Boys.
25. Mail Day.
26. Town Pump.
27. Picket Guard.
28. Going for the Cows.
29. The Bushwacker.
30. The Village Schoolmaster.
31. The Checker Player.
32. The Sharpshooters.
33. Checkers up at the Farm.
34. Weighing the Baby.
35. The Shaugran and Tatters.
36. The Tap at the Window.
37. The Mock Trial.
38. George Washington.

Choice Statuary by Thorwaldsen and other Eminent Artists.

39. Night, by Thorwaldsen.
40. Morning, "
41. Spring, "
42. Summer, "
43. Autumn, "
44. Winter, "
45. Apollo Belvidere.
46. The Three Graces.
47. The Greek Slave.
48. Eve before the Fall.
49. Night, by Copeland.
50. Morning, "
51. Ariadne and the Tiger.
52. Flora.
53. The Serenade.
54. The Courtship.
55. Burd Family Monument.
56. Simply to thy Cross I Cling.

Series XVIII.—Choice Dissolving Views.

PER SLIDE, $1.50.

1. *a* Voyage of Life, Childhood.   *Cole.*
  b Voyage of Life, Youth.   *Cole.*
  c Voyage of Life, Manhood.   *Cole*
  d Voyage of Life, Old Age.   *Cole.*
2. *a* The Seasons, Spring.
  b The Seasons, Summer.
  c The Seasons, Autumn.
  d The Seasons, Winter.
3. *a* Life's Day, Morning.   *Bellows.*
  b Life's Day, Noon.   *Bellows.*
  c Life's Day, Night.   *Bellows.*
4. *a* Prodigal Son. The Swineherd. *Dubufe.*
  b Prodigal Son. The Carousal. *Dubufe.*
  c Prodigal Son. The Return. *Dubufe.*
5. *a* The Christian Graces.   *Hicks.*
  b Il Penserosa.   *Hicks.*
  c L'Allegro.   *Hicks.*
6. *a* The Contraband.
  b The Recruit.
  c The Veteran.
7. *a* Steamboat Race on the Mississippi. Wooding up
  b Steamboat Race on the Mississippi. The Start
  c Steamboat Race on the Mississiupi The Explosion.

8. *a* Courtship for Second Wife. The Proposal.
  b Courtship for Second Wife. Ghost of First Wife appears.
  c Courtship for Second Wife. Great Consternation.
9. *a* Love and Marriage. The First Meeting.
  b Love and Marriage. Five Minutes After.
  c Love and Marriage. Five Years After.
10. *a* Brave Drummer Boy and his Father. Both Enlist in Union Army.
  b Brave Drummer Boy and his Father. In Battle against the Rebels.
  c Brave Drummer Boy and his Father. Both Die upon the Battle-field.
11. *a* Heathen Chinee. Euchre.
  b Heathen Chinee. The Right Bower.
  c Heathen Chinee. Twenty-four Jacks.
12. *a* Frigid Zone.
  b Temperate Zone.
  c Torrid Zone.

No.
13. a Heart's Ease. *Baxter.*
    b Lilies. *Baxter.*
    c Nora. *Baxter.*
14. a Faith. *Palmer.*
    b Hope. *Palmer*
    c Immortality. *Palmer*
15. a The Friendly Meeting.
    b A Temperance Meeting.
16. a Fondly Gazing. *G Smith.*

THE DESTRUCTION OF POMPEII.
4 Slides, $7.50

a The Poet's House, Pompeii. The beginning of the eruption of Vesuvius.
b The Last Night in Pompeii. The inhabitants fleeing through the fiery storm.
c Flight of the Pompeiians. The Terrified Populace are climbing a hill in the foreground; Vesuvius in eruption, and Pompeii and Herculaneum on the great mountain side, appear in the distance.

THE FAIRY FOUNTAIN.
Two Slides, $5.75

a Beautiful Golden Fountain.
b The Water fall and ripple.

17. a Fondly Gazing.
    b Empty Cradle.
18. a Ecce Homo. *Reni.*
    b Mater Dolorosa. *Dolce.*
19. a Jerusalem in her Grandeur. *Selous.*
    b Jerusalem in her Decay. *Selous.*
20. a Christus Consolator. *A. Scheffer.*
    b Christus Remunerator. *A. Scheffer.*
21. a Study. *Holfeld.*
    b Prayer. *Holfeld.*
22. a Mother's Dream. *Brooks.*
    b Believer's Mission. *Brooks.*
23. a Wife's Prayer. *Brooks.*
    b Dream of Hope. *Brooks.*
24. a Aurora. *Hamon.*
    b Feeding the Bird. *Hamon.*
25. a Beatrice Cenci. *Beranger.*
    b Evangeline. *Beranger.*
26. a English Farmyard. *Herring.*
    b English Homestead. *Herring.*
27. a The Luncheon. *Brochart.*
    b The Good Friends. *Brochart.*
28. a War. *Landseer.*
    b Peace. *Landseer.*
29. a To the Rescue. *Landseer.*
    b Saved. *Landseer.* Dog and Child.
30 a War. *Dore.*
    b Peace. *Dore.*
31. a Alexander and Diogenes. *Landseer.*
    b Jack in Office. *Landseer.*

No.
32. a Distinguished Member of the Humane Society. *Bateman.*
    b Nothing Venture, Nothing Have. *Bateman.*
33. a Lily of Ghent. *Absolon.*
    b Water Lilies. *Bouvier.*
34. a Cinderella. *Lejeune.*
    b Blue Bird. *Lejeune.*
35. a Mamma's Birthday. *Dobson.*
    b Remembrance. *Dobson.*
36. a At her Toilet. *Zuber Buhler.*
    b The Little Gourmand. *Zuber Buhler.*
37. a Liberality of Roman Women. *Coomans.*
    b Cornelia and her Jewels. *Schopin.*
38. a The Abduction. *Barrias.*
    b Vengeance. *Vernet.*
39. a The Lake. *Brochart.*
    b The Glacier. *Brochart.* Very choice.
40. a Cattle at Watering Place. *R. Bonheur.*
    b Sheep in Pasture. *R. Bonheur.*
41. a The Mother's Joy. *Amberg.*
    b The Widow's Comfort. *Amberg.*
42. a Morning Prayer. *Meyer von Bremen.*
    b Evening Prayer. *Meyer Von Bremen.*
43. a Saturday Night. *Absolon.*
    b Sunday Morning. *Absolon.*
44. a Going to the Club.
    b Returning from the Club.
45. a Going against the Stream. *Jenkins.*
    b Going with the Stream. *Jenkins.*
46. a High Life. *Landseer.*
    b Low Life. *Landseer.*
47. a Aspiring to Heaven. *Zuber Buhler.*
    b Regretting the Earth. *Zuber Buhler.*
48. a The Temperance Meeting. *Herring.*
    b The Friendly Meal. *Herring.*
49. a My First Sermon. *Millais.*
    b My Second Sermon. *Millais.*
50. a By the Seaside. *Brochart.*
    b Near the Falls. *Brochart.*
51. a Joy.
    b Sorrow.
52 a Fairy Tales.
    b Reading the Psalm.
53. a Grand Canal, Venice. Day.
    b " " " Moonlight.
54. a The Evening Prayer. *Frere.*
    b The Morning Kiss. *Frere.*
55. a The Onconvanience of Single Life.
    b The Real Convanience. (Comic.)
56. a The Quay at Liverpool. Outward Bound.

No.

56. *b* The Dock at Boston. (Comic.) Homeward Bound.
57. *a* The Settlement in the Backwoods. The Beginning.
 b The Settlement in the Backwoods. The Increase.
58. *a* Castle of Chillon, Lake Geneva, Switzerland. Day.
 b Castle of Chillon. Moonlight. Winter.
59. *a* Windsor Castle. Day.
 b Windsor Castle. Moonlight.
60. *a* Castle of Drachenfels. Summer.
 b Castle of Drachenfels. Winter Night.
61. *a* Castle of Ehrenfels on Rhine. Summer.
 b Castle of Ehrenfels on Rhine. Winter.
62. *a* Conway Castle. England. Day.
 b " " " Moonlight.
63. *a* Isola Bella, Italy. Day.
 b " " " Moonlight.
64. *a* Summer Landscape.
 b Winter "
65. *a* Prize Fighter Pet of the Fancy.
 b Exquisite. Pet of the Ladies.
66. *a* Grace before Meat.
 b Grace after Meat.
67. *a* Temptation.
 b Perdition.
68. *a* The Man. Temperance.
 b The Animal. Intemperance.
69. *a* Death-bed of the Righteous. John Wesley Praying.
 b Death-bed of the Wicked. Cardinal Richelieu Playing Cards.
70. *a* Abel's Sacrifice Received.
 b Cain's Sacrifice Rejected.
71. *a* Noah building the Ark.
 b Noah receiving Advice from above.
72. *a* Noah's Sacrifice.
 b " " Appearance of the Rainbow.
73. *a* Israelites passing through the Red Sea.
 b Destruction of Pharaoh and his Host.
74. *a* The Witch of Endor visited by Saul.
 b The Witch of Endor raising Samuel
75. *a* Flowers. Dahlias and Roses.
 b " Asters and Poppies.
76. *a* Fruits. Grapes.
 b " Currants.
77. *a* Before the Proclamation. A sad Negro Face.
 b After the Proclamation. A merry Negro Face.

No.

78. *a* The Schoolboy's First Cigar. Very manly.
 b The Schoolboy's First Cigar. Very sick.
79. *a* Good-night in Wreath of Flowers.
 b " Moonlit Sky.
80. *a* English Landscape. Tempest. Lightning.
 b English Landscape. Rainbow.
81. *a* The Repentant Sinner.
 b Knocking at the Gate.
 c Led by Jesus.
 d The Shores of the Beautiful River.
82. *a* Steamboat Race on the Mississippi. The Race.
 b Steamboat Race on the Mississippi. The Explosion.
83. *a* Fire in Philadelphia. The Alarm.
 b " " The Flames.
84. *a* The Protecting Scout. The Defenseless Woman.
 b The Protecting Scout. Appearance of Scout.
85. *a* Attack of the Monster. The Wicked Flea.
 b Attack of the Monster. The Boarding-house Bed-bug.
86. *a* Fops of the Past and Present. The Pre-historic Fop, according to Darwin.
 b Fops of the Past and Present. The Modern Fop, according to the 15th Amendment.
87. *a* A Ghostly Adventure. A Moonlight Reverie.
 b A Ghostly Adventure. A Ghostly Problem.
 c A Ghostly Adventure. The Problem Solved.
88. *a* American Meccas. Independence Hall, 1776.
 b American Meccas. U. S. Capitol, 1876.
89. *a* The Chinese Question. The Rivals.
 b " " Question Settled.
90. *a* Death of Sardanapalus. *Schopin.*
 b Socrates instructing Alcibiades. *Schopin.*
91. *a* House of a Poet. Last Days of Pompeii. *Coomans.*
 b Destruction of Pompeii. *Schopin.*
92. *a* How Jones became a Mason. Starting for the Lodge.
 b How Jones became a Mason. The Oath of Secrecy.
 c How Jones became a Mason. Riding the Goat.
 d How Jones became a Mason Jones has become a Mason.

76

No.
93. a Temperance—the Man.
  b Intemperance—the Animal.
94. a The Palace of Hell, built by Momus and his Crew.
  b Satan and his Chief Demons in Council.
95. a Going to Lodge. Leaves his Wife.
  b Returning from Lodge. Meets his Wife.
96. a Roman Chariot Race. The Departure.
  b Roman Chariot Race. The Triumph.
97. a The Coliseum's Martyrs. Wild Beasts and Victims. *Day.*
  b The Coliseum's Martyrs. Spirits hovering over the martyred forms in the deserted arena. *A splendid night effect.*
98. a French Wedding Party.
  b French Baptism Party.
99. a The Fountain of Love. *Brochart.*
  b Cupid a Captive. *Brochart.*
100. a The Age of Gold. *Schopin.*
  b The Paradise of Mahomet. *Schopin.*
101. a Hovering Angels. The Child Asleep.
  b Hovering Angels. The Cherub Choir. *Reynolds.*
102. a The Two Baths. Laughing Boy with Dripping Head.
  b The Two Baths. Laughing Girl Ducking a Kitten.
103. a The Lion and Lamb of the Golden Age.
  b The Lion and Lamb of the Present Age.
104. a The Jig: "St. Patrick's Day in the morning."
  b The Melody. "The Angel's Whisper."
105. a The Sculpture Gallery. Ancient Rome. *Tadema.*
  b The Picture Gallery. Ancient Rome. *Tadema.*
106. a How happy could I be with either, were t'other dear charmer away. *Oliver.*
  b Best be off with the old love, before you are on with the new. *Oliver.*

No.
107 a The Appian Way. Ancient Rome. Boulanger. [*Boulanger.*
  b Ladies' Apartment. Ancient Rome.
108. a The Promised Land. *Schopin.*
  b A Dream of Happiness. *Papety.*
109. a Fête at Court of Cleopatra. *Grolleau.*
  b Reunion at House of Aspasia. *Grolleau.*
110. a Christmas Night Without.
  b Christmas Night Within.
111. a Penna. Railroad Station, Broad St., Philadelphia. Exterior.
  b Pennsylvania Railroad Station, Broad St. Interior.
112. a Dance of the Veil. *Richter.*
  b Toilette of the Favorite. *Richter.*
113. a American Farmyard. Summer.
  b American Farmyard. Winter.
114. a Mammoth Cave, Kentucky. Red Fire Effect. [Fire Effect.
  b Mammoth Cave, Kentucky. Green
115. a Sunset Clouds.
  b Evening Clouds.
116. a The Smile. *Webster.*
Full well they laughed, with counterfeited glee,
At all his jokes, for many a joke had he.
  b The Frown. *Webster.*
Full well the busy whisper circling round
Conveyed the dismal tidings when he frowned.
117. a The Chorister Boys.
  b The Foundling Girls.
118. a "Tramp, tramp, tramp."
  In the prison cell I sit,
  Thinking Mother dear of you.
  b Tramp, tramp, tramp.
  The Boys came marching.
  Like a grand, majestic sea.
119. a Village Church on a Summer Morning.
  b Village Church on Christmas Eve.
120. a Toll Demanded.
  b Toll Taken.
121. a Joan of Arc finding the Sword.
  b She makes a sortie.
  c She suffers martyrdom.
122. a An Affair of Honor Two women fighting a duel.
  b Reconciliation.
123. a American Landscape. Winter.
  b American Landscape. Summer.

77

Series XIX.—Superior Colored Dissolving Views.

Producing Very Fine Dissolving Effects and Requiring the use of Two Lanterns.

No.   PRICE.

1. No CROSS, No CROWN. Four Slides, ... $6 00
   a Christiana gazing over the Sands of Time.
   b Christiana beholds the Cross of Christ.
   c Christiana dreams of the Beautiful Shore.
   d Christiana is Crowned by an Angel of Light.

2. ROCK OF AGES. Four Slides, ... 6 00
   a An angry Sea Swallowing a Wreck.
   b The Cross—The Rock of Ages—rises above the waters.
   c Faith, clinging to the Cross, is lifted above the waves.
   d Faith wings her flight heavenward.

3. THE GHOST SCENE. Three Slides for Dissolving Lanterns, . . 6 50
   A graveyard is represented; Day, Night, and Effects: the covering of one of the vaults is raised by a skeleton, which appears; other skeletons appear at different graves.

4. ORIGIN OF THE MOSS ROSE. Four Slides. With Poem, . . 6 00

5. ANGEL OF PEACE. Four Slides, ... 6 00
   a The Mother gazes fondly on her Babe.
   b The Mother sits beside an empty Cradle.
   c A Starry Sky above a Sleeping City.
   d The Angel of Death bears the Child heavenward.

6. ANGEL OF PEACE. Two Slides, ... 3 00
   a The city lies beneath wrapped in slumber, and scarcely discernible by the light of the new moon.
   b The Angel of Death, with outspread wings, flies across the scene, bearing the spirit of a child.

7. MERCY'S DREAM. Two Slides, ... 3 00
   a A beautiful woman is sleeping beneath a wide spread tree.
   b The vision of an Angel bearing a Crown of Light appears above her.

8. MOTHER'S GRAVE. Two Slides, ... 3 00
   a Three children are engaged in placing floral tributes upon their Mother's Grave.
   b Their Mother's Spirit descends and hovers over them.

9. BEETHOVEN'S DREAM. Two Slides, ... 3 00
   a The great Composer has fallen asleep at his piano.
   b The spirit of Music floats above him.

10. MARTYRED CHRISTIANS. Two Slides, ... 3 00
    a The lifeless figure of a beautiful woman floats upon the midnight waters.
    b Her Spirit is borne by Angels.

11. ORPHAN'S DREAM. Two Slides, ... 3 00
    a Tired of play, the Orphan Boy has fallen asleep.
    b His Mother's Spirit appears, bending lovingly over him.

12. SHIPWRECKED MARINERS. Two Slides, ... 3 00
    a Two mariners cast upon a rocky coast, discover a ship in the distance at day-dawn.
    b Morning advances, and the ship approaches.

No.		Price
13. Abou Ben Adhem. Two Slides. With Poem,		$3 00
a The first appearance of the Angel.		
b The second appearance of the Angel.		
14. Star of Bethlehem. Two Slides,		3 00
a Wise Men of the East journeying toward Jerusalem.		
b The Son of Man appears in a radiant light.		
15. Vase of Growing Flowers. Two Slides,		3 00
a The Flowers in Bud.		
b The Flowers in full Bloom.		
16. Falls of Niagara. Two Slides,		3 00
a General View of Falls in Summer		
b A beautiful Rainbow appears in the Mist.		
17. Highlander's Dream of Home. Two Slides,		3 00
a A Highland Soldier asleep by his Camp fire.		
b Vision of Home appears above the fire.		
18. Rock of Ages. Two Slides,		3 00
a The Cross—the Rock of Ages—rises above an Angry Sea.		
b Faith clinging to the Cross, is saved from perishing.		
19. Birth of Venus. Two Slides,		3 00
a Flying Cupids announce the coming of Venus.		
b The beauteous Venus is born of the Ocean's foam.		
20. Washington's Tomb. Two Slides,		3 00
a Tomb of Washington at Mount Vernon, on the Potomac.		
b The Spirit of Washington appears within the Tomb.		
21. Washington's Dream. Two Slides,		3 00
a Falls asleep over his war map at Valley Forge, Penna.		
b Beholds a vision of America's future prosperity.		
22. American Soldier's Dream of Home. Two Slides,		3 00
a Asleep by the Camp fire.		
b A vision of Home appears in the smoke of the fire.		
23. Napoleon. Two Slides,		3 00
a Powerful at the head of his army.		
b Powerless on the barren rock at St. Helena.		
24. White and Red Roses. Two Slides,		3 00
a White Rose, emblematic of Purity.		
b Red Rose and Cupid with Bow, emblems of Love.		
25. The Bachelor's Reverie. Two Slides,		3 00
a The Bachelor indulges in a twilight reverie.		
b A vision of his first love appears.		
26. Mosque of Omar. Two Slides,		3 00
a Mosque of Omar, Jerusalem, by day.		
b The Mosque illuminated by night.		
27. Salisbury Cathedral, England. Two Slides,		3 00
a The beautiful Cathedral by day.		
b The illuminated Cathedral by moonlight.		
28. Westminster Abbey, London. Two Slides,		3 00
a The magnificent Abbey by daylight.		
b The illuminated Abbey by moonlight.		
29. Tower of London. Three Slides,		4 50
a The Tower by day.		
b The Tower by night.		
c The Great Fire destroying the Tower.		

No.   Price

30. FAUST AND MARGUERITE. Two Slides, . . . $3 00
    a Faust in his Laboratory tempted by Mephistopheles.
    b Flames dart from Mephistopheles' lamp, and vision of Marguerite appears.

31. LOOK NOT UPON THE WINE WHEN IT IS RED. (Very good.) Two Slides, 3 00
    a A beautiful Girl in the abandon of the dance, wine cup in hand.
    b A hideous Skeleton continues the dance, a serpent creeping from the cup. (Proverbs xxiii, 31)

32. GOOD MORNING! Two Slides, . . . 3 00
    a The Window of a Palatial Mansion, with shutters closed.
    b Shutters fly open and reveal a fair face and figure.

33. A DREAM OF IMMORTALITY. Two Slides, . 3 00
    a A beautiful Lady lies dreaming upon a couch.
    b Angels place a Crown upon the sleeper's brow.

34. THE PROTECTING SCOUT. Two Slides, ... 3 00
    a A defenseless Woman and Children attacked by Indians.
    b Appearance of the Protecting Scout.

35. THE WOOD-NYMPH'S BATH. Two Slides, . 3 00
    a An embowered lake in the forest, by moonlight.
    b A Wood-nymph, upon a couch of lilies, floats upon the waters.

36. THE HANDWRITING ON THE WALL. Two Slides, ... 3 00
    a Belshazzar in the midst of a Bacchanalian Revel.
    b Daniel reads the words, "*Mene, Mene, Tekel, Upharsin.*"

37. THE FLIGHT OF AURORA AND HER TRAIN Two Slides, . 3 00
    a A gorgeous mass of rosy clouds.
    b Aurora, followed by Apollo, and a host of Goddesses.

38. THE LITTLE FOXES' RETREAT. Two Slides, . . . 3 00
    a Trunk of an old hollow tree, in which is a large hole.
    b Three saucy-looking little Foxes peep out of the hole.

Series XX.—Dissolving Views with Magnificent Movable Effects.

Two Lanterns are Required for the Exhibition of these Slides.

1. WATER-MILL IN PENNSYLVANIA. Four Slides, . . $10 00
    a A summer day; the water-wheel in motion.
    b The moon rises and produces a rippling effect on the water.
    c The Mill in winter; the ground covered with snow.
    d Snow-storm; the white flakes fall thick and fast.

2. FORT SUMTER, CHARLESTON HARBOR. Four Slides, . 7 50
    a The Fort by daylight in time of Peace.
    b The Fort by moonlight in time of Peace.
    c On fire during Bombardment.
    d Fire and smoke curl upward from the Fort.

3. BAY OF NAPLES AND MOUNT VESUVIUS. Three Slides, . . 6 00
    a Grand Panorama by daylight.
    b Night; the mountain in eruption.
    c Fire and smoke rise from the burning crater.

4. CASTLE OF ST. ANGELO AND CHURCH OF ST. PETER, ROME. Three Slides, 6 00
    a The Church and Castle by daylight.
    b Gorgeous illumination on Easter night.
    c Fireworks fly through the heavens.

No.   Price.

5. LIFE NEAR THE NORTH POLE. Three Slides,   . $3 00
   a The Arctic Regions by day.
   b Night among the icebergs.
   c Brilliant Aurora Borealis flashes upward in the northern sky.

6. MOUNT ÆTNA, ISLAND OF SICILY. Three Slides, .   .   . 6 00
   a The great mountain by day
   b Volcanic eruption at night.
   c Fire and smoke pour from the flaming cone.

8. MAGICIAN AND CAULDRON. Two Slides, ... 6 50
   a A weird incantation scene. A Magician is standing within his cave,
     waving a wand over a bubbling cauldron.
   b Ghosts, Witches, Imps, Gnomes, etc., fly from the cauldron.

10. NAIAD QUEEN OF THE RIVER RHINE. Two Slides,   .   . 4 00
   a Moonbeams glisten on the Rhine, upon whose shore a castle rises in
     frowning outline.
   b The Naiad Queen appears seated upon her Throne of Shell, and
     glides over the waters playing her wonderful harp.

11. EXPRESS TRAIN, ... 3 00
   a A railroad bridge by moonlight with a forest in the background.
   b A Locomotive and Train of Cars dash by, the headlight and sparks
     making a brilliant effect.

12. THE SERENADE IN VENICE. Two Slides,   .   .   . 4 00
   a Grand Canal by moonlight. Castle in the foreground.
   b A Venetian Cavalier approaches in a Gondola and sings before the
     Castle. A lady appears upon the balcony above him.

13. STEAMER LEAVING PORT. Two Slides,   .   .   . 4 00
   a A vast harbor, and city in the distance.
   b A Steamer glides across the harbor and puts to sea.

14. FIRE IN PHILADELPHIA. Two Slides,   .   .   . 4 00
   a Street by night. Fire over the housetops. The alarm.
   b A Steam Fire Engine dashes by, drawn by two prancing horses.

15. LAKES OF KILLARNEY, IRELAND. Two Slides,   4 50
   a Angels fold their wings and rest
     In that Eden of the West,
     Beauty's Home, Killarney.
   b Moon rises, and the waters ripple.

16. MARTYRED CHRISTIAN. Two Slides,   .   .   . 4 50
   a The body of a beautiful woman floats upon the moonlit waters.
   b Her Spirit is borne upward by Angels. (Beautiful effect.)

17. MAGIC LILY. Two Slides, ... 4 00
   a The Beautiful Lily of the East, the home of Fairies.
   b A Fairy with a golden wand rises from the bosom of the Lily.

18. HAUNTED ABBEY. Two Slides, ... 4 00
   a Tomb in the ruins of an old English Abbey.
   b A Ghost rises from the Tomb.

19. THE SKELETON DANCE IN KIRK ALLOWAY. Two Slides,   .   . 6 25
   a Ruins of Kirk Alloway, Scotland, scene of Tam O'Shanter's vision.
   b A Skeleton executes a fantastic dance among the ruins.

20. WATER-MILL IN THE ALPS. Two Slides,   .   .   . 4 50
   a Summer in the Alps; revolving water-wheel.
   b Winter; snow-clad mountains; wheel frozen fast.

21. HOLLAND WIND-MILL. Two Slides,   .   .   . 4 50
   a A Dutch Wind-mill by moonlight.
   b Daylight; the fans of the Wind-mill revolving.

Series XXI.—Choice Colored Slides with Motion.

These may be used in a Single Lantern, and produce very beautiful an' new effects by giving revolving, slip, or eccentric motions to the scene.

No. 1.

NO.		PRICE.
1	THE DANCING SKELETON, a new and very striking effect, . . .	$4 50
	By a peculiar Mechanical arrangement, the skeleton is made to bow with his head in his hand, to dance, etc., in a most LIFE-LIKE manner. Music may be used, and the figure be made to dance to jig-time with the most mirth-provoking effect.	
2.	CASTLE ON LAKE MAGGIORE, ITALY. Very beautiful, ...	3 50
	An Italian night scene. A boat containing a Lover glides over the moonlit waters; he serenades his mistress in the Castle; after which she glides out upon the balcony.	
3.	BOMBARDMENT OF FORT SUMTER; The Ironsides throwing shell, .	3 50
4.	VIEW OF OLD RUINS, which, by being turned around, changes to Portrait of an Old Woman, ...	3 00
5.	HOLLAND WIND-MILL, with Revolving Fans, ...	3 00
6.	FOUNTAIN, ...	4 50
7.	NEWTON'S DISK, Revolving Slides, with prismatic colors, for re-composing white light. Beautiful effect, ...	3 50
8.	RATCATCHER—Man sleeping—Awakes and swallows one rat after another in quick succession. Very laughable, ...	4 00
9.	MOUNT VESUVIUS—Eruption—Throws out Fire and Smoke · Good for one lantern, ...	4 50
10.	MOVING WATERS. Represents the Waters moving in the Moonlight. A very beautiful and natural effect, ...	2 25
11.	GOOD NIGHT IN WREATH. A Wreath of Flowers, in which appears a Young Girl holding a Candle. She disappears, and is succeeded by the words Good Night. Excellent for closing, ...	3 00
14.	THE DANCING SAILOR, a new and striking effect, having a motion similar to the Dancing Skeleton, ...	4 50
15.	SWISS WATER-MILL; Wheel Revolves, ...	2 50
16.	THE AQUARIUM, in which fish move about, ...	4 50
17.	THE BEEHIVE, surrounded by Flying Bees. Very fine, . . .	4 50
18.	CURTAIN SLIDE. One Slide, ...	2 50
	Represents the Rolling up of a Curtain, and discloses a very pleasing landscape for commencing a Dissolving-view Exhibition.	
19.	GYMNAST, ...	4 50
	Best, ...	6 00
20.	MOON SLIDE. One Slide, Lever Movement, ...	2 25
21.	ROCKET SLIDE. May be used with any night scene, causing rocket to rise and explode in the air, producing a shower of colored stars falling down, ...	3 00
22.	AURORA BOREALIS for use with arctic series, ...	2 50

New Effect Slides for a Single Lantern.

$2.25 EACH.

These are quite New and Beautifully Colored.

1. FINGAL'S CAVE. Interior seen by a torch and partly by daylight; a brilliant red light gradually illuminates the scene.
2. RACE-COURSE AT MUNICH. The familiar scene of the race-course, with crowds of spectators, are here exhibited, when the running horses pass rapidly across.
3. BLUE GROTTO AT CAPRI ISLAND. The interior is seen by dim daylight coming through the entrance; a boat with torches comes across and illuminates the grotto, which, after the boat passes, appears again in blue, as at first seen.
4. RAILROAD TRAIN. A viaduct near Verviers, Prussia, by moonlight; a train with lighted-up cars passes across.
5. BURNING HOUSE. A fire-engine, drawn by two galloping horses and a lot of firemen, passes across rapidly.
6. SUNSET ON THE ALPS. The Glacier of Loffel-Spitz is seen in full daylight; evening approaches, and the shepherd is returning over the bridge across the limpid brook to his cottage when the sunset glow on the icy peaks illumes the scene.
7. THE NEW BRIDGE ACROSS THE ST. GOTHARD. A beautiful day-scene, representing the usual Alpine travel across.
8. CASTLE STOLPEN, IN GERMANY. An evening scene; the moon rises and lights up by degrees the castle and landscape.
9. VENICE. View of the Grand Canal at evening—illuminated gondolas pass along the waters.
10. BALLOON ASCENSION. A large number of spectators are collected on a public square surrounded by houses; the balloon is already inflated and anchored in the middle; at the signal the balloon begins to rise, and passes out of sight at last.
11. VIEW OF THE "SIEBENGEBIRGE" (SEVEN HILLS), RHINE. The scene is from the Terrace of Castle Ehrenfels, across the placid waters of the Rhine; the windows of the castle are lighted; a steamboat passes along.
12. POLAR SCENERY. Showing a beautiful Aurora, with changing colors.
13. LIGHT-HOUSE ON THE COAST. A steamer is passing.

Extra Fine Dissolving Views, $3.50 the set.

FOR DOUBLE LANTERN.

14. THE BLUE GROTTO OF CAPRI. Artificial illumination—visitors with torches passing in boats.

15. NIAGARA. Horseshoe Falls; day scene; a rainbow appears in the spray.

16. THE THORSTEIN. A natural bridge in; day scene changing to moonlight.

17. CASTLE EBERSTEIN IN BAVARIA. Day and night.

18. STOLZENFELS ON THE RHINE. Day and night.

New and Beautiful
Dissolving * Views,
FINELY PAINTED.

Each set is composed of several pictures, provided with mechanism for producing various interesting effects.

---

1. **Viaduct Across the Loire in France** [3 views and 3 changes].

A windmill with set of revolving fans is seen in the distance through the arches; evening comes on, and the lamps on the bridge are lighted one by one; a train with lighted-up cars passes across, ... $10 00

2. **The Ill-Fated Steamer** [4 views].

A vessel starts with a fair wind; a storm comes up; the ship is struck by lightning and takes fire, which is seen curling up; the wreck is finally left by the crew, which is saved on a raft, ... 10 00

3. **The Forge in the Alps** [4 views].

A beautiful mountain scene in full sunlight; an approaching storm obscures the sky; a flash of lightning strikes the house, and flames are seen issuing and increasing; the burning ruins are seen at night by moonlight, ... 10 00

4. **A Swiss Mountain Scene** [5 views], with a cascade of running water.

A gang of brigands are seen passing a small bridge and entering a hut standing by the roadside, where presently a light appears, ... 10 00

5. **A View of the Cathedral of Trieste** [5 views].

On the plaza in the foreground a fountain of sparkling water, which at nightfall is illuminated with a variety of colors, ... 10 00

6. Villa Ostia, Italy, on the Tiber [3 views].

The villa is illuminated at night and rockets are sent up from different
points, which explode and display a shower of colored stars falling
down, and are reflected in the water, ... $7 50

**7. Castle Hohenschwangau in the Tyrolese mountains [3
views].**

A sunset view ; at evening the castle is illuminated, and a fountain in the
foreground plays in all colors, ...   7 50

**8. Tropical Landscape, the sources of the Jumna, India [3
views].**

A waterfall is rolling down a precipitous rock forming a pool of water
below ; an elephant is enjoying a bath, ...   7 50

9. An Avalanche in the Alps [3 views].

A Swiss cottage is tumbled down the mountain by falling snow and rocks ;
the moon lights up the place lately occupied by the cottage, ...   7 50

10. The Bay of Naples, seen from Virgil's Tomb [3 views].

Mount Vesuvius is seen in the distance with a column of smoke issuing from its crater ; night comes on and the volcano becomes active,
and emits fire and lava, ...   7 50

**11. Palace Fredericsberg, near Copenhagen, Denmark
[3 views].**

The palace is illuminated at night ; at times the moon emerges from
the passing clouds and the light is reflected from the ripples of the
water, ...   7 50

12. The Wetterhorn, Swiss [3 views].

A very characteristic landscape in full sunlight ; evening comes on and
darkens the scene ; soon after sunset the snowcapped mountain becomes
brilliant by the Alpine glow ; the moon now appears in the sky, and
lights in the cottages, ...   7 50

13. Ruins of Pompeii [3 views].

A magnificent painting ; in the distance Vesuvius towers skyward ; the
day is past, and the scene changes, and Vesuvius hurls fire and lava
high into the air, ...   7 50

14. The Imperial Palace, St. Petersburg, by day.

The palace is illuminated at night, and sleighs pass in front, ...   7 50

15. Emperor Fountain at Aix la Chapelle.

Palace illuminated at night ; fountain plays all colors. 7 50

Series XXVI.—Colored Photographic Comic Slides, with Movable Slip.

PRICE, $1.00 EACH.

No.
1. Backing out of Going to Market.
2. Spring and Fall.
3. Race Go-as-you-please.
4. Clearing the Letter-box.
5. Skipping Girl.
6. Boy and Donkey.
7. Dancing Imp.
8. Dancing Negro.
9. Good-night.
10. Man Beating his Donkey.
11. Irishman Driving Pig.
12. Boy Falling off Pig.
13. Man Asleep Swallows Mice.
14. Organgrinder and Monkey.
15. Man with Growing Tongue.
16. Performing Elephant.
17. Monkey Dipping Cat.
18. Boy Chasing Butterfly.

No.
19. Boy Firing off Cannon.
20. Wizard Raises Demon
21. Dog Catches Monkey (Tail comes off
22. Monkey Teasing Cat.
23. A Pear—a Pair.
24. The End of the Tale (Tail).
25. Dentist.
26. Punch Bowl.
27. Man Shaving and Cat.
28. Frog Jumps out of Pie.
29. Blacksmith at Work.
30. Cow Tossing Dog.
31. Monkey Teasing Old Woman.
32. Lady on Kicking Mule.
33. Elephant and Keeper.
34. Boy Teasing Dog.
35. Boy on Two Stools.

Series XXVIII.—Chromatropes, or Artificial Fire-works.

1. "THE NATIONAL FLAG" CHROMATROPE. Each, ... $3 00
From designs expressly made to introduce the colors of our glorious National Flag. We have five different patterns of this Chromatrope.

2. "THE GEOMETRICAL" CHROMATROPE. Each, ... 3 00
A variety of entirely new and original patterns, of superior Chromatic and Geometrical effects. We have many different styles of this Chromatrope.

3. "THE WASHINGTON" CHROMATROPE. Each, ... 3 75
A new and beautiful design, with a Photographic Likeness of Washington in the centre (copied from Stuart's celebrated painting in the Boston Athenæum), and the Stars and Stripes revolving around it in glorious array.

Special Selection of Transparencies

from the Works of Ganot's Physics, Guillemin's Forces of Nature, Pepper's Playbook of Science, Science for All, etc.

ELECTRICITY.

Iron filings attracted by bar magnet.
Magnet formed by two compound bars.
Natural magnet and its armature.
Horseshoe magnet with armature and keeper.
Magnetizing horseshoe magnet.
Magnetic pendulum.
Magnetic needle showing inclination and declination.
Electricity excited by friction.
Electrophorus.
Electricity excited by influence.
Electrical induction through a series of conductors.
Electric flow.
Gold leaf electroscope.
Gold-leaf electroscope in use.
Coulomb's electrometer.
Otto von Guericke's machine.
Ramsden's plate machine.
Holtz machine.
Carré's di-electrical machine.
Armstrong's hydro-electrical machine.
Electric sparks.
Electric chimes.
Luminous tube.
Volta's pistol.
Charging Leyden jar.
Discharging Leyden jar.
Battery of electrical jars.
Universal discharger.
Experiment of perforating glass.
Leyden jar with movable coatings.
Franklin's kite experiment.
Different kinds of lightning.
Flash of lightning photographed.
Lightning conductor.
Limit of protection by conductors.

MAGNETISM.

Voltaic element.
Voltaic pile.
Electricity developed by chemical action.
Daniell's cell.
Grove's cell.
Bunsen's cell.
Leclanche's cell.
Decomposition of water by Voltaic battery.
Faraday's experiment in electrolysis.
Oersted's experiment.
Ampère's law.
Schweigger's multiplier.
Astatic galvanometer.
Ohm's law.
Resistancepile.
Wheatstone's bridge.
Voltaic element and galvanometer showing current.
Action of magnet on a current.
Rotation of current by magnet.
Action of current on a solenoid.
Rhumkorff's coil and interrupter.
Rhumkorff's induction coil.
Stöhrer's induction coil.
Gramm's mag-electro machine.
Construction of Gramm's armature.
Diagram of apparatus in Morse telegraphy.
Hughes' printing telegraph.
Diagram of Hughes' printing telegraph.
Electric bell.
Jablochkoffe electric candle and lantern.
Foucault's regulator for electric light
Browning's regulator for electric light
Image of the carbon point.
Magnetic curves from catalogue, p. 130.

MECHANICS.

Action and reaction.
Parallelogram of forces.
Composition and resolution of forces.
Parallel forces, the arithmetical lever.
Equilibrium of two forces.
The pulley
The compound pulley.
Simple levers.
Wheel and axle.
Windlass.
Inclined plane.
The screw.
The wedge.

Elasticity—spring balances.
Plumb-line, vertical to fluid surface.
Centre of gravity.
Different positions of equilibrium.
Stable and unstable equilibrium.
Attwood's machine.
Diagram of motion of pendulum.
Pendulum, verification of law.
Application of pendulum to clock.
Tisley's compound pendulum.
Tisley's compound curves.
The Vernier.
Cathetometer.

HYDROSTATICS.

Equality of liquid pressure.
Hydraulic press.
Hydrostatic paradox.
Barker's mill.
Equilibrium of liquids.
Equilibrium of liquid in communicating
    vessel.
Artificial fountain.

Artesian wells.
Principle of Archimedes, demonstrated.
Hydrostatic balance, sp. gr. of solid.
Hydrostatic balance, sp. gr. of fluid.
Specific gravity bottle.
Nicholson's hydrometer.
Cohesion of liquids, dew drops.
Capillary elevation and depression.

PNEUMATICS.

Illustration of atmospheric pressure.
Madgeburg hemispheres.
Torricellian experiment.
Simple form of barometer.
Wheel barometer.
Aneroid barometer.
Air pump.
Mercury pump.

Condensing pump.
Fountain in vacuo.
Hero's fountain.
The siphon.
The suction pump.
The forcing pump.
Centrifugal pump.
Screw turbine.

HEAT.

Simple pyrometer.
Unequal expansion of different metals.
Compensating pendulum.
Pendulum with compensating bars.
Expansion of liquids and gases by heat.
Filling a mercurial thermometer.
Determination of freezing point.
Determination of boiling point.
Thermometer scales.
Maximum and minimum thermometer.
Maximum density of water.
Cold produced by expansion of gases.
Carré's ice machine.

Ebullition; water singing.
Ebullition; water boiling.
Comparative volumes of water and steam.
Construction of thermopile.
Thermopile and galvanometer.
Radiation of heat in straight lines.
Radiation of heat; law of inverse squares.
Reflection of radiant heat.
Burning mirror.
Refraction of heat; burning glass.
Absorption of radiant heat by air.
Modern locomotive engine, section.
Gas engine, Otto.

SOUND.

Propagation of pulse or wave.
Propagation of sound from a bell to ear.
Propagation of sound-wave in a tube.
Speaking-tube, mouth-piece, and whistle.
Invisible woman.
Reflexion of sound.
Reflexion of echo.
Refraction of sound.
Refraction of sound by a sound lens.
Velocity of sound in air.
Velocity of sound in water.
Savarte-toothed wheel experiment.
Seebeck's syren.
Vibrations of tuning-fork.
Lissajou's method of showing same.

Helmholtz double syren.
Vibrations and nodes in a spring
Melde's experiment; string attached to tuning fork.
Chladny's sand figures.
Sand figures on membranes.
Propagation and reflection of liquid waves on the surfaces of an elliptical bath of mercury.
Diagram of vibrating tuning forks in state of coincidence and interference.
Mammetric flames.
Mammetric flames, note and octave, note and third.

TELEPHONE, MICROPHONE, AND PHONOGRAPH.

Various uses of the telephone.
Simple electro-magnet.
Bell's articulating telephone.
Bell's articulating telephone in circuit.
Bell's articulating telephone in use.
Bell's telephone, various forms.
Proprietor communicating with factory.
Telephone applied to warfare.
Edison's carbon telephone.
Edison's shouting telephone, principle.

Hughes' pencil microphone.
Hughes' pencil microphone in use.
Watch ticking heard by microphone.
Edison's phonograph.
Preice & Tyndall working the phonograph.
Phonograph with governor in use.
Gramme's magneto-electric machine.
Gramme's magneto-electric distributor.
Brush magneto-electric distributor.

MAGNETIC CURVES.

Photographs from the Actual Figures made with Iron Filings by Professor S. P. Thompson, D. Sc.

Magnetic curves of bar magnet.
Magnetic curves of horseshoe magnet.
Magnetic lines of force of single pole.
Curves of attraction of two magnets.
Curves of repulsion of two magnets.
Two parallel magnets attracting.
Two parallel magnets repelling.
Lines of force of dissimilar poles.
Lines of force of similar poles.
Horizontal section of electro-magnet.
Lines of force of electro-magnet.
Action of magnetic field on a small magnet.
Circular lines of force round a galvanic current.
Lines of force of current in horizontal wire.
Magnetic field of a looped conductor.
Field of two parallel attracting currents.
Field of two parallel repelling currents.
Attraction of two parallel horizontal currents.
Repulsion of two opposed parallel currents

Magnetic field of oblique currents.
Field of horizontal and vertical currents.
Lines of force of current deflecting a needle.
Stable position of needle near vertical currents.
Neutral position of needle near vertical current.
Unstable position of needle near vertical current.
Field of force of a galvanometer.
Field of magnetic needle in a circuit.
Field of magnet attracting current.
Attraction of North Pole into a simple circuit.
Repulsion of South Pole out of a simple circuit.
Mutual rotation of current and magnet pole.
Spiral field of magnet rotated by current running through it.
Converse spiral field of South Pole rotated by current running through it.

PHYSICAL GEOGRAPHY.

SCENES AND VIEWS.

1 Plain and Table-land.
2 Hills, Mountains.
3 Valleys, Cañons.
4 Glaciers, Icebergs.
5 Sea or Lake.
6. Rivers, Water Falls.
7 Waterspout.
8 Cyclones, Sandstorm.
9 Avalanche.
10 Mirage.
11 Magnetic Chart lines of equal variations.

MAPS AND CHARTS.

12 Isobars for January.
13 "   "   July.
14 Isothonical lines for January.
15 Isothonical lines for July.
16 Relief map of North America.
17 "   " South America.
18 "   " the United States.
19 "   " Europe.
20 "   " Africa.
21 "   " Asia.
22 "   " Australia.

PHYSICAL GEOLOGY.

Lecture By the REV. H. N. HUTCHINSON, B.A., F. R. G. S.

1 The Seasons.
2 Circulation of Atmosphere.
3 Waterspout.
4 July Isobars and Prevailing Winds.
5 January Isobars.
6 Classes of Clouds.
7 Distribution of Rainfall.
8 Formation of Caverns in Limestone.
9 Stalactites. Jenolan Caves.
10 Grotto. Jenolan Caves.
11 Artesian Well.
12 Cañon, Colorado.
13 Vishnu's Temple (Horizontal Strata).
14 Pinnacle in Kanab Cañon.
15 Niagara, Horseshoe Falls.
16 Vermillion Cliffs, Utah.
17 Vermillion Cliffs, Section.
18 Inversnaid Falls.
19 Loch Lomond.
20 Monte Rosa.
21 Rhone Glacier.
22 Viesch Glacier.
23 Glacier and Crevasse.
24 Iceberg.
25 Greenland Glacier.
26 Stacks of Duncansby.
27 The Matterhorn.

28 Section across the Alps.
29 Concentric Earthquake Waves.
30 Gifu (Japan), after the Earthquake.
31 Bridge Ruined by Earthquake, Gifu.
32 Earthquake Effects at Diana Marina.
33 Vesuvius in Eruption (1872).
34 Ideal Section of Volcano.
35 Vesuvius, Crater and Lava Stream.
36 Vesuvius. Lava Stream.
37 Fingal's Cave, Staffa.
38 Basaltic Columns, Giant's Causeway.
39 Grand Geyser, Yellowstone Park.
40 Foraminifera (Depth 1850 Fathoms).
41 Globigerina Bulloides.
42 Foraminifera (Depth 2900 Fathoms).
43 Orbulina.
44 Foraminifera (Magnified).
45 Polycistina (Bermuda).
46 Holtenia.
47 Ventriculites.
48 Depths of the Sea (Atlantic Ocean).
49 Depths of the Sea (Pacific Ocean).
50 Coral Formations.

LIGHT AND OPTICS.

REFRACTION OF LIGHT.

1. Linear propagation of light.
2. The prism—deviation of light.
3. Decomposition of light by prism, solar spectrum.
4. Unequal refrangibility of rays of light.
5. Formation of images by small apertures.
6. Experiments illustrating refraction.
7. Forms of lenses.
8. Fresnet's lens for light-house.
9. Formation of images by convex lenses.
10. Reverted image of landscape.
11. Intensity of illumination—law of inverse squares.
12. Index of refraction.
13. Chromatic aberration.
14. Achromatic prism and lens.
15. Spherical aberration.
16. Atmospheric refraction.
17. Effect of refraction on sun-set.
18. Diogram to explain wave-lengths.
19. Diagram to interference of waves.
20. Table of wave-lengths.
21. Double refraction Iceland spar.
22. Nicol's prism and Polariscope.
23. Colored rings in crystals with Polariscope.
24. Recomposition of light by 7 mirrors.
25. Recomposition of complimentary colors by polariscope.
26. Reflection of light by plane mirror.
27. Illustration of law of reflection.
28. Total reflection—limiting angle.
29. Phenomena of total reflection.
30. Reflection from concave mirrors.
31. Foci of concave mirrors.
32. Virtual and real image.
33. Scattering light by irregular surfaces.
34. The rainbow.
35. Explanation of same.
36. Mirage.
37. Explanation of Pepper's Ghost.

Spectrum Analysis.

PLAIN PHOTOS, 50 CENTS PER SLIDE.

1. Sectional view of a Spectroscope.
2. Increased Dispersion by a series of Prisms.
3. Automatic arrangements for a Battery of Prisms.
4. Necessity for the use of a narrow Slit.
5. Slit and Comparison Prism.
6. Use of Collimating lens.
7. Large Spectroscope with returning ray.
8. Solar Spectroscope. (Secchi.)
9. Star Spectroscope. (Secchi.)
10. Apparatus for the Spectra of Metallic Vapors.
11. Solar Spectrum, photographed by Draper.
12. Chart of Radiation Spectra. No. 1.
13. Chart of Absorption Spectra. No. 1.
14. Reversal of Sodium Lines on Screen.
15. Long and short Lines in Spectra, with explanation.
16. Long and short Lines in Solar Prominences.

A Set of Slides, Illustrating Spectrum Analysis. Finely Colored.

PER SLIDE, $1.50.

No.
1. Decomposition of Light by Prism (Solar Spectrum)
2. Comparative Intensity of Heating, Luminous, and Chemically Active Rays (in Spectrum).
3. Fraunhofer's Map of Solar Spectrum (1814–15).
4. The Spectroscope.
5. Spectra of the Sun, Beta Cygni and Hydrogen.
6. Spectra of Potassium, Rubidium, Sodium, and Lithium.

No.
7. Spectra of Carbon, Comet II, Brorsen's Comet (1868), Spark and Nebulæ.
8. Spectra of Aldebaran and Alpha Orionis.
9. Spectra of Chlorophyll, Chloride of Uranium, Magenta, and Blood.
10. Gassiott's Spectroscope, made by Browning.
11. Coincidence of Spectrum of Iron with 65 of the Fraunhofer Lines.
12. Spectra of Sun, Chromosphere, Prominences, and Corona.
13. The Atmospheric Lines.

Geography.

UNCOLORED, 50 CENTS EACH.

1. Map of World, on Mercator's projection.
2. North America.
3. South America.
4. Europe.
5. Asia.
6. Africa.
7. Oceanica.

Set of Twenty Anatomy and Physiology.
Beautifully Painted Pictures.

PER SLIDE, $1.50.

1. Human Skeleton.
2. " Skull.
3. Section of the Spine, etc.
4. Teeth, and Structure of same.
5. Muscles, Front View.
6. " Back "
7. " of the Head, Neck, and Face.
8. General View of the Digestive Organs, in place.
9. The Digestive Organs.
10. The Stomach, Liver, and Pancreas.
11. The Thoracic Duct.
12. Heart and Lungs.
13. Diagram of Circulation.
14. Skin, and Structure of same.
15. Brain and Spinal Cord.
16. General View of the Nerves
17. Fifth Pair of Nerves
18. Facial Nerves.
19. Diagram of the Eye
20. Anatomy of the Ear.

A Set of Twenty on Microscopic Anatomy.

THE HEART AND HOW IT BEATS.

With Reading.

Human Physiology Popularly Explained; or, the House We Live In.

By Mr. William Furneaux,

Department of Astronomy

Movable Astronomical Diagrams.
The Motion Produced by Rack-work.
PACKED IN A BOX.

1. The Solar System, showing the Revolution of all the Planets, with their Satellites, round the Sun.
2. The Earth's Annual Motion round the Sun, showing the Parallelism of its axis, thus producing the Seasons.
3. The cause of Spring and Neap Tides, and the Moon's phases during its revolution.
4. The Apparent Direct and Retrograde Motions of Venus or Mercury, and its Stationary Appearance.
5. The Earth's Rotundity, proved by a Ship sailing round the Globe, and a line drawn from the eye of an observer placed on an eminence.
6. The Eccentric Revolution of a Comet round the Sun, and the appearance of its tail at different points of its Orbit.
7. The Diurnal Motion of the Earth, showing the Rising and Setting of the Sun, illustrating the cause of Day and Night, by the earth's rotation upon its Axis.
8. The Annual Motion of the Earth round the Sun, with the Monthly Lunations of the Moon.
9. The various Eclipses of the Sun, with the transit of Venus ; the Sun appears as seen through a telescope.
10. The various Eclipses of the Moon ; the Moon appears as seen through a telescope.

The set of 2½ inch diameter, with Lecture,   .   . $35 00
"   "   3   "   "   "   .   .   45 00

Astronomical Diagrams.

From Asa Smith's Atlas.

PLAIN PHOTOGRAPHS, 50 CENTS EACH.

No.
1. Vienna Observatory
2. Terrestrial and Celestial Globes.
3. The Solar System.
4. Signs of the Zodiac.
5. The Earth and Seasons—Solstices and Equinoxes.
6. Phases of the Moon.
7. Tides and inclination of Moon's Orbit.
8. Moon's Nodes and Eclipses.

No.
9. Eclipses of Sun and Moon.
10. Direct and Retrograde Motion of Planets.
11. Orbits of Asteroids, Mars and Jupiter.
12. " Saturn and position of rings.
13. " Uranus and Neptune.
14. Refraction, Parallax.
15. Kepler's Laws.

A. THE SUN.

1. *Granular appearance* of the surface under highest powers. *C. Flammarion.*
2. Sun spot, projecting radiations, penumbra feebly developed. *C. Flammarion.*
3. Solar willow leaves, after Nasmyth. *Secchi.*
4. Sun spot. *Langley.*
5. Sun spot, } *Harvard*, { 1872.
6. " " } { 1873.
7. Photosphere and sun spot. *Flammarion.*
8. Partial view of sun disc during eclipse of May, 1882.
9. Solar prominence, 1872. *Harvard Univ.*
10. " " 1873. "
11. A solar explosion. *C. Flammarion*, 1882.
12. Total Eclipse, August 7th, 1869. *Harvard Univ.*
13. " " December 1st, 1871. Corona, photosphere, and prominences.
14. Comparative size of a prominence and the earth.
15. Panoramic view of a partial solar eclipse, May, 1882.
16. Wolfe's sun spot numbers.

B. THE MOON.

1. Map of Full Moon and Key to Craters.
2. Full Moon, photograph by Rutherford.
3. First Quarter, " "
4. Last " "
5. Mare Crisium at Full Moon. *Piazzi Smith.*
6. Lunar Craters—A. Linni, Gassendi, Plato, Aristarchus, Herodotus, etc.
    B. Cassini, Plinius, Arzachel, etc.
    C. Hypatin, Guttenberg, Julius Cæsar, etc.
    D. Torricelli, Agrippa, Godin, etc.
7. Ideal Lunar landscape. *C. Flammarion.*
8. Earth seen from the Moon.
9. Extinct Craters on the Earth, in Iceland.
10. " " " " in Auvergne, France.
11. An Eclipse of the Moon seen from the Earth.
12. Telescopic view of partial Eclipse of Moon.

C. THE SOLAR SYSTEM.

No.
1. Solar System (scale of 1 millimeter to ten million Fr. miles). *C. Flammarion, 1881*
2. Annual elliptic orbit of the earth round the Sun, " "
3. Comparative inclination of axes of Earth and Venus, " "
4. " " " " Jupiter, " "
5. " sizes of Earth, Mars, and Mercury, " "
6. " " " and Jupiter, " "
7. " " " Uranus and Neptune, " "

TELESCOPIC APPEARANCE OF

9. Mercury, April 14th, 1871, a little more than half-illuminated—South Pole very bright. *Bothk. Obs.*
10. Mercury, April 22d, 1871, crescent shape; two dark spots near the edge—North Pole bright. *Bothk. Obs.*
11. Occultation of Venus by Moon. *C. Flammarion.*
12. Phases of Venus. *C. Flammarion.*
13. Mars, 1873. *Harvard Univ.*
14. " September 10th, 1877. *Potsdam Obs.*
15. Jupiter, 1873. *Harvard Obs.*
16. " red spot from royal Astro. notices, 1886.
17. Saturn in four positions of his rings.
18. " several divisions on rings. *C. Flammarion.*
19. " rings as seen from the Planet. *C. Flammarion.*
20. " 1872. *Harvard Univ.*
21. " ring system seen from above. *C. Flammarion.*
22. Comparative size of Sun and Planets.

D. STARS—NEBULÆ—AURORA, Etc.

1. Star Cluster in Hercules. *Harvard Univ.*
2. " Taurus.
4. " Perseus.
5. Ring Nebula in Lyra. *Harvard Univ.*
6. Dumb-bell Nebula. "
7. Andromeda " "
8. Trifid " "
9. Grand " in Orion "
10. " " central part, "
12. Spiral " in Canes Venat. *Secchi.*
14. Crab " in Taurus. "
16. The Pleiades.
18. Double and Multiple Stars.
23. Cluster in Greyhounds.
24. Nebulæ in Scutum. *Sobiesky.*

E. COMETS.

1. Great Comet of 1811, with naked eye. *Bruhn's Atlas.*
3. Halley's Comet, October 28th, 1835, with naked eye. *Bruhn's Atlas.*
5. " " " 8th, 1835, telescopic drawing by Bessel. *Bruhn's Atlas.*
7. Donati's " 1858, drawing by Bond *Secchi.*
8. " " October 5th, 1858, telescopic. *Bruhn's Atlas.*
10. " " September 30th, 1869, do. *Harvard Univ.*
13. Coggia's " June 10th, 1874. "
14. " " July 13th, 1874. "

F. STAR MAPS.

No.
2. Northern Celestial Hemisphere. *Kendall.*
4. Southern " " "
6. Constellations, visible January 21st to April 17th.
8. " " April 18th to July 21st.
10. " " July 22d to October 31st.
12. " " November 1st to January 20th.

DIAGRAMS OF THE CONSTELLATIONS.

1. Scorpion.	9. Eagle, Antinous.	16. Andromeda.
2. Lion.	10. Orion.	17. Perseus.
3. Ophiuchus.	11. " Trapezium.	18. Coma Berenice.
4. Hercules, Crown	12. Dragon, Little Bear.	19. Fishes.
5. Bull.	13. Virgin, Crow.	20. Twins.
6. Swan.	14. Pegasus.	21. Scorpion.
7. Aquarius, Goat.	15. Cassiopeia.	
8. Great Bear, Greyhounds.		

ASTRONOMICAL PHOTOGRAPHS TAKEN AT THE LICK OBSERVATORY.
Published by kind permission of Prof. Holden.

1. View of the Lick Observatory.
2. 36-inch Equatorial of the Lick Observatory.
3. Eye-end of the 36-inch Equatorial Telescope.
4. Micrometer of the 36-inch Telescope.
5. Spectroscope on the 36-inch Telescope.
6. The Sun, showing Sun Spots and Faculæ, August 28th, 1893, reduced from Negative taken with the Photoheliograph.
7. The Sun, showing Sun Spots and Faculæ, August 30th, 1893, reduced from Negative taken with the Photoheliograph.
8. The Sun, showing Sun Spots and Faculæ, September 3rd, 1893, reduced from Negative taken with the Photoheliograph.
9. The Sun, showing Sun Spots and Faculæ, September 4th, 1893, reduced from Negative taken with the Photoheliograph.
(Nos. 6 to 9 show the same group of spots in its passage across the disc).
10. Portion of the Sun's Disc, showing Groups of Spots; Contact Print from Negative taken with the Photoheliograph, May 15th 1894.
11. Portion of the Sun's Disc, showing Groups of Spots; Contact Print from Negative taken with the Photoheliograph, June 18th, 1894.
12. Portion of the Sun's Disc, showing Groups of Spots Contact Print from Negative taken with the Photoheliograph, June 19th, 1894.
13. Portion of the Sun's Disc, showing Groups of Spots; Contact Print from Negative taken with the Photoheliograph, June 19th, 1894, enlargement of the principal group of spots shown on No. 12.
14. Portion of the Sun's Disc, showing Groups of Spots; Contact Print from Negative taken with the Photoheliograph, June 20th, 1894.
(Nos. 11 to 14 show the same groups of spots).
15. Total Solar Eclipse of April 16th, 1893. Eight pictures of Solar Corona taken with the Photoheliograph Lens of 40 feet Focus, by J. M. Schaeberle, at Mina Bronces, Chile (Lick Observatory Expedition).
16. Total Solar Eclipse of April 16th, 1893; taken with the 40 feet Photoheliograph by J. M. Schaeberle, at Mina Bronces, Chile. Exposure, 25 seconds.
17. Total Solar Eclipse of April 16th, 1893; taken with the 40 feet Photoheliograph by J. M. Schaeberle, at Mina Bronces, Chile. Exposure, 8 seconds.
18. Total Solar Eclipse of April 16th, 1893; taken with the 40 feet Photoheliograph by J. M. Schaeberle, at Mina Bronces, Chile. Exposure, 16 seconds.
19. Photograph of the same Eclipse taken with the Dallmeyer Lens of the Lick Observatory, by J. M. Schaeberle.
20. Portion of the Sun's Limb during the same Eclipse, showing Solar prominences and lower portion of the Corona. Contact Print from the large Negative taken with the 40 feet Photoheliograph, by J. M. Schaeberle.
21. Portion of the Sun's Limb during the same Eclipse, showing Solar prominences and lower portion of the Corona. Contact Print from the large Negative taken with the 40 feet Photoheliograph, by J. M. Schaeberle.
22. Portion of the Sun's Limb during the same Eclipse, showing Solar prominences and lower portion of the Corona. Contact Print from the large Negative taken with the 40 feet Photoheliograph, by J. M. Schaeberle.
23. Portion of the Sun's Limb during the same Eclipse, showing Solar prominences and lower portion of the Corona. Contact Print from the large Negative taken with the 40 feet Photoheliograph, by J. M. Schaeberle.
24. The Moon, taken 1890, July, 20d. 7h. 52m., Moon's Age, 4d. 3h.
25. The Moon, taken 1890, Nov., 16d. 5h. 57m., Moon's Age, 4d. 12h.
26. The Moon, taken 1893, Nov., 15d. 7h. 6m., Moon's Age, 7d. 14h.
27. The Moon, taken 1893, July, 21d. 9h. 1m., Moon's Age, 8d. 16h.
28. The Moon, taken 1891, Oct., 11d. 7h. 32m., Moon's Age, 9d. 2½h.
29. The Moon, taken 1892, Mar., 8d. 13h. 16m., Moon's Age, 10d. 5½h.
30. The Moon, taken 1890, Oct., 26d. 10h. 19m., Moon's Age, 13d. 1h.
31. The Moon, taken 1890, Aug., 4d. 12h. 39m., Moon's Age, 19d. 8h.
32. The Moon, taken 1890, Nov., 3d. 13h. 58m., Moon's Age, 21d. 5h.
33. Jupiter, taken with enlarging Lens in 36 inch Telescope.
34. Jupiter, taken with enlarging Lens in 36 inch Telescope. (Sept. 26th, 1892).
35. Nubecula Major, or Greater Magellanic Cloud; reduced from Negative taken with the Dallmeyer Lens by J. M. Schaeberle, April 9th, 1893; 3 hours Exposure.
36. Clusters in Perseus; reduced from Negative taken with the 36-inch Telescope

Supplementary List.

STARS.

Algol during an eclipse.
Proper motion of the stars, south of 70° north polar distance.
    "   "   "   north 75°   "   "   "
The Nebulæ and Clusters of Sir J. Herschel's catalogue.
Epilon Lyræ and comps.
Herschel's star gauges in the southern heavens.
Cluster in Perseus, direct photograph. *Paris Obs.*, 1884.
Star Sphere of Flaristed's projection of all stars visible to the naked eye
The Pleiades, direct photograph by the *Henry Bros.*, Paris.
Great Nebula in Pleiades, direct photograph, by *Isaac Roberts*, 1889.
χ The γ Argus region and neighboring Clusters, S. Milky way, photo.
χα and β Crucis and Coal-sack region.   "   "
Multiple Stars drawn to scale by *G. F. Chamber.*
χ Region of Milky way about a Cygni. *Dr. Max Wolf,* photo.
Original design of the nebulosity of Merops. *M. Temple.*
The Trapezium of Orion with the 36″ refractor.
Milky way in Sagittarius, direct photograph. *H. C. Russel.*
Belt and Sword of Orion,   "   "   "
Apparent Orbit of ζ Hercules,
    "   "   ς Sagittarii.
    "   "   ς Ursæ Majoris.
    "   "   70 Ophinchi.
    "   "   Satellite of Sirins.
Cluster 2 M Aquarii—*Lord Ross.*

SPECTROSCOPY.

Stellar Spectra of 3d class. *Duner.*
    "   "   4th   "   "
Spectrum of Hydrogen in White Star, compared with Spectrum of the Orion nebula, *Huggins.*
Photographic Spectrum of Grand Nebula, Orion. *Huggins.*
Spectrum of Mira, showing lines of Hyd. and Calcium. *Pickering.*
Spectrum of Uranus. *J. E. Keeler.*
Spectra of gaseous stars. *Vogel.*
Spectrum of Botalquesa (a Orionis) *Harvard College.*
Changes in Spectrum of Nova Cygni. *Vogel.*
Comet Spectra.
Secchi's types of Stellar Spectra.

.

PLANETS.

Transit of Venus, 1871, ingress and egress. *R. A. S*
Chart of planet Venus, by *Nusten*, 1881–1890.
Venus in 1881, by *W. F. Denning.*
   "   " 1890, by *Perrotin.*
Phases of Venus. *J. V. Schiapparetti*
Plane map of Mars. "   "
    "   "   "   "
Dark lines on "   "   "
Telescopic view "   "   "
Orbits of the Satellites of Mars.
Telescopic view of Jupiter, March 5th, 1886. *Dearborn Obs.*
   "   "   " November 24th, 1883, and January 22d, 1884.
   "   "   "   " 7th, 1884, and February 27th, 1885
   "   "   " August 28th, 1890. *Lick Obs.*
Occultation of   moons.
Jupiter's red spot, 1880.
Mercury in September, 1889. *Tacchini.*
Shadow of Saturn on the rings at different seasons of its year.
Planet's Orbit inside of Saturn orbit.

NEBULÆ.

Photo. Neb. in Andromeda. *Roberts.*

COMETARY.

Orbit of great Comet of 1882.
Orbits of meteoric swarms associated with comets.
Close coincidence of cometary orbits within the earth's orbit.
Diagram illustrating the influence of Jupiter on comet.
Coggins' Comet from July 1st to July 14th, with 10" refractor.
a. The great Comet of 1882 in 26" Equatorial, October.
b. "   "   "   "   " November.
The "Sheath" and the attendant of Comet of 1882.
Donati's Comet passing Arcturus, 1858. (No. 7, old list in catalogue.)

MISCELLANEOUS.

The Ptolómaic system.
August and November meteors.
Determination of distance of a planet from the Sun.
Inversion of the orbit of a Satellite.

www.ingramcontent.com/pod-product-compliance
Lightning Source LLC
Chambersburg PA
CBHW022004190326
41519CB00010B/1376